Addition Facts

in five minutes a day!

Susan C. Anthony

Instructional Resources Company
P.O. Box 111704
Anchorage, AK 99511-1704

Teach with less effort, more SUCCESS!

Instructional Resources Company is a teacher-owned business which aims to assist teachers in maintaining high objective standards while helping *all* students reach them. Our materials and workshops are designed to save teachers time while helping them nurture the excitement of learning in their students, build confidence through early success, and build a framework of background information to which new learning can be connected throughout life. We welcome feedback from anyone using our materials with students!

Instructional Resources Company
P.O. Box 111704
Anchorage, AK 99511-1704
(907) 345-6689

Instructional Resources Company
3235 Garland
Wheat Ridge, CO 80033

Books by Susan C. Anthony are listed below. Request a catalog for an up-to-date list.

Mathematics
Addition Facts in Five Minutes a Day
Subtraction Facts in Five Minutes a Day
Multiplication Facts in Five Minutes a Day
Division Facts in Five Minutes a Day
Casting Nines: A Quick Check for Math Computation

Reference
Facts Plus: An Almanac of Essential Information
Facts Plus Activity Book
Encyclopedia Activity for use with The World Book Encyclopedia

Spelling
Spelling Plus: 1000 Words toward Spelling Success
Dictation Resource Book for use with Spelling Plus
Homophones Resource Book
Personal Dictionary Masters

Handwriting
Manuscript Handwriting Masters
Cursive Handwriting Masters

ISBN 1-879478-23-4

CONTENTS

Why Memorize Math Facts?

There are many benefits to memorizing basic math facts. With the methods and materials in this book and its companion books, an investment of five minutes a day in grades 2 (addition), 3 (subtraction), 4 (multiplication) and 5 (division), will pay dividends for a lifetime, *conserving* time and effort in the long run. Here are some benefits of memorizing math facts:

- Knowing basic facts frees students' minds to concentrate fully on math concepts, skills, and problem-solving. Overlearned, mastered facts become background information, immediately available but requiring little or no conscious thought.
- Math is exact. A wrong answer because of a missed fact is no less wrong than a wrong answer because a concept isn't understood.
- After students have mastered the basic facts, teachers can progress much more quickly through the curriculum. Working with students who *don't* know their facts at higher grade levels places a drag on the process of teaching mathematical concepts and skills.
- Because the facts are so basic and so useful in math, students who know them have a much better foundation for junior high, high school, and even college than those who do not.
- Although calculators and computers are wonderful tools, a person who must locate a calculator in order to figure out 5 + 5 would not be considered mathematically literate by most, regardless of his/her understanding of concepts.
- Knowing the basic facts gives students a sense of self-esteem. Students who know the facts generally feel more confident of themselves and have a more positive attitude toward math than students who do not.
- Memorizing facts need not interfere in any way with teaching concepts, problem-solving, etc. It is a separate activity which requires five minutes a day in grades 2-5, leaving the rest of math time for teaching everything else.

Getting Started

Tips for Teachers Using This Book

- First read p. 2 for a quick explanation of how the five-minutes-a-day activity works and a step-by-step checklist for advance preparation.
- Scan pp. 3-6 for answers to any questions that may have arisen as you read p. 2.
- If you open the book to any page and have questions about what you see, page backwards to the section divider and read the pages immediately following it.
- Worksheets are on the front side of each page. Corresponding answer keys are on the back.
- Teachers are welcomed and encouraged to adapt these materials according to their own preferences and the needs of their students. For example, you may choose to have students pass the 10's worksheet before memorizing the 4's.

Timed Test Masters

TIMED TESTS

The black-line masters in this book are intended to be flexible so that teachers can use them in any way that is appropriate for their students. In this section are ideas and suggestions based on how *I* use these materials. Adapt as necessary.

Basic Procedure for Timed Tests

1) Students get out their math folders.
2) Worksheets are distributed.
3) I say, "Ready, set, GO!" and start a stopwatch.
4) Students work through the test as quickly and accurately as they can.
5) As each student finishes, a note is made of his/her time.
6) After five minutes, I say, "STOP!"
7) I collect papers from students who have finished in three minutes or less and check them myself. Anyone with 100 facts correct in three minutes or less "passes."
8) Other students check their own tests and record their scores and times in their folders.

Step-by-Step Checklist for Preparing to Use this Basic Procedure

- ❐ Read the introductory information on pp. 2-6. Highlight key points for easy reference later!
- ❐ Label a file folder for each of the timed tests on pp. 7-49.
- ❐ Reproduce numerous copies of each of the timed tests and place them in the appropriate folders. For a classroom, at least 100 copies of each test will likely be needed. To conserve paper, ditto on the back of waste paper or make two-sided copies, then recycle.
- ❐ Prepare a folder for each student. Instructions and reproducible masters are on pp. 52-56. Staple a sticker sheet on the left (p. 54) and either the Record of Timed tests or a graph on the right (pp. 55-56; instructions for preparation before duplicating are on p. 52).
- ❐ Reproduce answer keys for each student on colored paper. Answer keys are on the backs of each test master, pp. 8-50. Staple packets of keys together and place in folders.
- ❐ Prepare or purchase a chart of addition facts to post in back of the classroom,
or reproduce the chart of facts on p. 53 and tape it in each folder
or have students fill in the chart of addition facts on p. 85 and tape it in the folder.
- ❐ Read the third paragraph on p. 4 and decide how you want to keep times. Get a stop watch **or** prepare a time chart as shown on p. 53 and get a pointer.
- ❐ Buy stickers to award after individual tests are passed and colorful certificates to acknowledge students who have memorized *all* addition facts to mastery. Decide what other special reward you might offer when mastery is accomplished.
- ❐ Reproduce and cut apart numerous congratulations certificates (p. 58). Store them in an accessible place. These go home after individual tests are passed, with stickers.
- ❐ Optional: Reproduce personal flash cards for each student's use as sets of facts are introduced (pp. 61-82). These can be stored in plastic sandwich bags in the folders.
- ❐ Optional: Reproduce a classroom set of copies of "Race to the Finish" on p. 86 for a substitute to use if you are unexpectedly absent.

Addition Timed Tests: Introductory Lessons

Introduce a Set of Facts

- Use any method you prefer to have students generate sets of facts with manipulatives. Children should understand the concept of addition before committing the facts to memory.
- Distribute personal flash cards (pp. 61-82) and sandwich bags. Students may cut their own sets apart or cutting can be done beforehand. The bags are for storage.
- Have students practice with flash cards alone or with partners for several days before beginning timed tests on a set of facts.

The First Timed Test

- Explain the importance of memorizing basic math facts. See p. v for reasons it's important.
- Distribute the Addition 0-1 timed test (p. 7) to all students. Explain that they should do as much as they can in five minutes, but that you will stop them when five minutes is up. It is OK not to finish. Speed will come with practice and accuracy is more important than speed.
- Time students for five minutes.
- Collect and check the papers. This shows the teacher each individual's starting point.

Keeping Scores and Times

- Distribute another copy of the 0-1 timed test to all students the next day. Explain to students how you'll keep time. See the third paragraph on p. 4 and the chart on p. 53 for options. If *most* students have great difficulty completing 100 facts in five minutes, adjust your mastery standard, perhaps to four minutes. Later in the school year, try again for three minutes.
- Time students for five minutes and have them write down their times on the worksheet.
- Give each student a copy of the answer key. Show them how to check their own papers and record their scores, then collect and recheck the papers. Work with students on checking their own papers until they are able to do it accurately.

Folders and Individualized Progress

- When you feel students are ready (perhaps in a few weeks), distribute folders and have students write their names on them. The folders will allow students to proceed at their own speed through the tests. Include copies of all answer keys in the folders to facilitate self-checking. Collect and check folders and papers only from those students who complete a test in three minutes or less. Begin work in folders with the test you've been doing as a group.

Additional tips

- Timed tests are always the first activity in my math class. Students learn to have folders out.

- The first timed test in addition is zeros and ones (p. 7). Some students may pass the first day in three minutes or less. I make a big deal out of this, and emphasize that they've already mastered 21 of the 66 basic addition facts (considering 5 + 3 and 3 + 5 as the same fact). There are only 45 to go and this is just the first *day*. We have a whole *year* to learn the rest. Although other facts will be more difficult to memorize than zeros and ones, this is an excellent start! Certificates go home, there's applause, stickers, stars, etc. This **early success** gives students confidence and enhances self-esteem.

- For those who don't pass the first day: Don't worry. You'll have another chance tomorrow and the next day and the next. This is not a race. You have a whole *year* to learn these facts. It is naturally easier for some kids than for others to memorize these because some people are lucky to be born with mathematical talent. If math isn't your area of talent, you have *other* areas. Learning the facts might take you longer than someone else, but it doesn't matter how long it takes, it just matters that you *learn* them by the end of the year. If you do, you'll be *way ahead* of most students in third grade around the United States. There are a lot of *junior high and high school* students who don't know their facts well. It might be harder for you than for someone else, but *you can do it!* Persistence will pay off and the more difficult it is for you, the more credit you'll deserve when you accomplish it!

- Once students have their folders, they can progress at their own speed. Eventually, each student will be working at a level which challenges him/her. To distribute the worksheets, I walk around the room with a stack of file folders containing blank worksheets and ask, "Who needs 3's? Who needs 0-3's?" and so on. This takes only a few minutes.

- Recording the times can be done in any of several ways. I sit at my desk with a stopwatch and scan the room. As soon as a student finishes, he raises his hand and I record his initials and the time. I read these aloud as soon as the five minutes are up and students note their own times in their folders. Another method is to make a large time chart as shown on p. 53. The teacher stands next to the chart and points to each time as it elapses on the stopwatch or a wall clock. When a student is finished, he looks up and records the time to which the teacher is pointing. Another possibility is to have students use their own stopwatches.

- I have students do all their own checking and record-keeping until they complete 100 facts in three minutes or less. When they do, I collect the entire folder including the test. There are generally only a few of these each day. I check these tests, and if all 100 facts are correct, I record the date on a chart such as that on p. 57, give a sticker (p. 54), and prepare a certificate to go home (p. 58). If any facts were missed, the test must be tried again the next day.

- Once a student has passed the Addition 10 worksheet on p. 39, there are five different worksheets with all addition facts mixed. I pass out packets of these to save time.

- When a student passes a final worksheet (Addition I, II, III, IV or V on pp. 41-50), there is a BIG reward. I give a colorful certificate with a real Susan B. Anthony dollar. This is *well worth the cost!* Some students spend their dollar and others keep it as a souvenir. On a schoolwide basis, there can be recognition at an awards assembly or a mention in the school newspaper. Emphasize to students who haven't yet passed that their time will come! Their reward is ready and waiting for them and you have confidence that they'll get it!

- After a student passes the final worksheet, she can continue doing the five final worksheets for faster times, or go on to practice subtraction. To maintain mastery, even children who have passed addition should do at least one worksheet per week throughout the year. The goal is *overlearning* for mastery. To motivate students who have finished to improve their speed, I tell them that one day late in the year we'll invite the principal to do a timed test with us to see if any students can beat her. Since she'll be "out of practice" there is a possibility that one or more students will be able to complete the worksheet in less time.

Questions and Answers

- **What about students who can't handle time pressure or have test anxiety?** Occasionally, early in the year, a student will panic and "freeze" during a timed test. If this happens, I counsel with the student. I tell him that even though we call this a "test" it is not graded and there is no way to fail. If he got only ten answers correct today and can get eleven correct tomorrow, that is progress and will eventually lead to his knowing all the facts. There is no hurry. We have all year. As long as he's headed in the right direction, he can relax and just do his best. Some parents have told me that their children initially felt threatened by timed tests but got over it after a short time and began to like them. In the process, they lost some of their fear of tests in other subject areas.

- **What about the aspect of competition?** I post a wall chart such as that shown on p. 57. Children know where they stand in relation to others. I *continually* emphasize that learning math facts is not a race, that it is easier for some people because they were lucky enough to be born with talent in this area, and that the goal is for *everyone* to do 100 facts correctly in three minutes or less by the end of the *year.* In my experience, the top students do race and compete with each other. That is one reason they like the tests and aren't bored by them. I encourage students to support and cheer for each other, and they do. When a student who's had difficulty passes a test, he receives more cheers and applause than the top students! When parents come to the classroom, kids at all levels often escort them directly to the wall chart to show how they're doing. This program will work with or without a wall chart, but in my experience, it can be handled so that students benefit from whatever competition naturally takes place.

- **Does this activity ever get boring?** It is a routine, rather like brushing one's teeth or having lunch at a particular time. In my experience, students do not become bored, possibly because as soon as they master one worksheet, they move on to another. Those who have passed *all* worksheets are given other options. In my experience, many of them choose to continue doing the timed tests even when they are free to choose to do something else.

- **What about special education students or others who just can't succeed at this?** I have had a few students who were truly unable to complete a worksheet in three minutes. Once they got behind and were not making progress, I allowed them to pass on to the next worksheet by getting 100 correct in *five* minutes or less. To get the final reward for knowing *all* the facts (a Susan B. Anthony dollar in my case), they had to pass in three minutes or less like everyone else. There were students who had a great deal of trouble early in the year, who went on the five minute plan and passed all of the worksheets, and who then worked on speed and passed the final test in three minutes for their reward. I escorted them straight to the principal with their papers. Parents were called immediately with the good news. I emphasized to the whole class that this had not been easy, that this child had persisted and succeeded despite the frustration he'd felt, that *anyone* can persist and succeed at things that are difficult for them. I emphasize to the child that he is now far *ahead* of many older students who don't know their facts. A *very* few students did not achieve the goal of 100 facts correct in three minutes or less during the year. However, *these students still knew the facts!* I had achieved *my* objective. Even students who find math difficult like the timed tests and are proud of the measurable progress they make.

- **What about students who can't write fast?** Students who are slow writers may pass the tests one-on-one orally with me when they are ready. Generally, I confer with parents beforehand.

- **What about left-handed students?** Left-handed students may be frustrated if their hand covers the question. Allow them to write the answers in the column to the *left* of the question.

- **What about counting on fingers?** Although students may count on their fingers, I emphasize that they have not *mastered* the facts if they depend on their fingers or a chart. The idea is to *memorize* the facts so that they move beyond the need to use anything other than their own heads.

- **Does being "behind" harm a student's self-esteem?** Although teachers may avoid comparing students, children already know that they are better in some subject areas than others. They already know who is doing better and worse in various subjects than they are. Achieving an objective goal such as getting 100 facts correct in three minutes, despite the frustration and discouragement that may occur along the way, is an excellent way to *enhance* self-esteem. Children learn that they can achieve things they may have initially thought impossible. They can achieve *anything* they put their mind to if they persist and believe in themselves.

- **What about the students who already know the facts?** Top students may proceed through the worksheets as quickly as one per day (finishing in as little as three weeks). They then work to improve their speed on the final tests, or proceed with learning the subtraction facts. Students who have passed all tests have the option of doing only one test per week and using the time on the other four days to read or do special activities. In my experience, they generally prefer to continue doing the tests for faster times. This is an area in which they shine and they like it!

- **What do students do when they finish early and are waiting for the five minutes to elapse?** They can go over their papers to make sure they didn't make any careless errors. They can use the key to check and score their paper. They can read or do a quiet activity. They can quietly encourage a nearby friend.

- **What about cheating?** In my experience, there has been little or no cheating on this activity. The teacher checks any papers which count for "passing" so there is little incentive to cheat. If cheating does occur, I have a talk with the student, expressing my disappointment and emphasizing that the goal is to learn the facts, not pass the tests. I offer to do whatever I can to help the child learn the facts. If the problem were to recur (it never has), parents would be called.

- **How can parents help?** I send home flashcards after we've used them in class, and at the first conference provide parents with a photocopied set of all the timed tests, which they can copy and use at home. Some children like to do a "practice test" every night at home!

- **What about days when a substitute is in the room?** Timed tests are somewhat difficult for a substitute to manage once students are progressing individually. If I know I will be gone, I pass out the appropriate worksheets the day before so that a substitute can just keep times and collect the folders of students who finish in three minutes or less. Sometimes I pull the worksheets each student needs the night before and write names on them so distribution is easy. Another option is to have a substitute do the "Race to the Finish" activity on p. 86. Warn students in advance not to expect regular timed tests when a substitute is teaching.

Addition 0-1

Name ______________________ Date ______________ Score ______ Time ______

(1)	7 + 0		(26)	0 + 10		(51)	9 + 0		(76)	0 + 4	
(2)	1 + 2		(27)	1 + 0		(52)	5 + 1		(77)	10 + 0	
(3)	3 + 0		(28)	3 + 1		(53)	1 + 7		(78)	1 + 8	
(4)	5 + 1		(29)	0 + 4		(54)	10 + 0		(79)	2 + 0	
(5)	1 + 9		(30)	1 + 3		(55)	0 + 8		(80)	7 + 1	
(6)	1 + 3		(31)	6 + 0		(56)	1 + 9		(81)	1 + 1	
(7)	0 + 2		(32)	0 + 7		(57)	6 + 1		(82)	0 + 0	
(8)	10 + 0		(33)	1 + 10		(58)	0 + 2		(83)	1 + 6	
(9)	6 + 1		(34)	0 + 0		(59)	4 + 0		(84)	8 + 0	
(10)	0 + 0		(35)	4 + 1		(60)	10 + 1		(85)	3 + 0	
(11)	1 + 6		(36)	0 + 9		(61)	1 + 8		(86)	5 + 1	
(12)	5 + 0		(37)	7 + 0		(62)	0 + 7		(87)	1 + 7	
(13)	0 + 8		(38)	8 + 1		(63)	1 + 0		(88)	1 + 2	
(14)	1 + 4		(39)	0 + 3		(64)	1 + 5		(89)	7 + 0	
(15)	7 + 1		(40)	3 + 1		(65)	0 + 10		(90)	0 + 1	
(16)	0 + 3		(41)	0 + 1		(66)	2 + 1		(91)	10 + 1	
(17)	9 + 1		(42)	3 + 0		(67)	6 + 0		(92)	1 + 4	
(18)	8 + 0		(43)	1 + 5		(68)	1 + 8		(93)	2 + 0	
(19)	1 + 10		(44)	9 + 1		(69)	4 + 0		(94)	5 + 0	
(20)	0 + 5		(45)	0 + 6		(70)	10 + 1		(95)	1 + 6	
(21)	4 + 1		(46)	1 + 1		(71)	0 + 6		(96)	1 + 0	
(22)	1 + 7		(47)	8 + 0		(72)	9 + 0		(97)	1 + 1	
(23)	2 + 0		(48)	0 + 5		(73)	0 + 4		(98)	9 + 0	
(24)	1 + 1		(49)	1 + 2		(74)	8 + 1		(99)	1 + 4	
(25)	0 + 1		(50)	7 + 1		(75)	1 + 1		(100)	2 + 1	

Addition 0-1

KEY

(1)	7 + 0	7
(2)	1 + 2	3
(3)	3 + 0	3
(4)	5 + 1	6
(5)	1 + 9	10
(6)	1 + 3	4
(7)	0 + 2	2
(8)	10 + 0	10
(9)	6 + 1	7
(10)	0 + 0	0
(11)	1 + 6	7
(12)	5 + 0	5
(13)	0 + 8	8
(14)	1 + 4	5
(15)	7 + 1	8
(16)	0 + 3	3
(17)	9 + 1	10
(18)	8 + 0	8
(19)	1 + 10	11
(20)	0 + 5	5
(21)	4 + 1	5
(22)	1 + 7	8
(23)	2 + 0	2
(24)	1 + 1	2
(25)	0 + 1	1
(26)	0 + 10	10
(27)	1 + 0	1
(28)	3 + 1	4
(29)	0 + 4	4
(30)	1 + 3	4
(31)	6 + 0	6
(32)	0 + 7	7
(33)	1 + 10	11
(34)	0 + 0	0
(35)	4 + 1	5
(36)	0 + 9	9
(37)	7 + 0	7
(38)	8 + 1	9
(39)	0 + 3	3
(40)	3 + 1	4
(41)	0 + 1	1
(42)	3 + 0	3
(43)	1 + 5	6
(44)	9 + 1	10
(45)	0 + 6	6
(46)	1 + 1	2
(47)	8 + 0	8
(48)	0 + 5	5
(49)	1 + 2	3
(50)	7 + 1	8
(51)	9 + 0	9
(52)	5 + 1	6
(53)	1 + 7	8
(54)	10 + 0	10
(55)	0 + 8	8
(56)	1 + 9	10
(57)	6 + 1	7
(58)	0 + 2	2
(59)	4 + 0	4
(60)	10 + 1	11
(61)	1 + 8	9
(62)	0 + 7	7
(63)	1 + 0	1
(64)	1 + 5	6
(65)	0 + 10	10
(66)	2 + 1	3
(67)	6 + 0	6
(68)	1 + 8	9
(69)	4 + 0	4
(70)	10 + 1	11
(71)	0 + 6	6
(72)	9 + 0	9
(73)	0 + 4	4
(74)	8 + 1	9
(75)	1 + 1	2
(76)	0 + 4	4
(77)	10 + 0	10
(78)	1 + 8	9
(79)	2 + 0	2
(80)	7 + 1	8
(81)	1 + 1	2
(82)	0 + 0	0
(83)	1 + 6	7
(84)	8 + 0	8
(85)	3 + 0	3
(86)	5 + 1	6
(87)	1 + 7	8
(88)	1 + 2	3
(89)	7 + 0	7
(90)	0 + 1	1
(91)	10 + 1	11
(92)	1 + 4	5
(93)	2 + 0	2
(94)	5 + 0	5
(95)	1 + 6	7
(96)	1 + 0	1
(97)	1 + 1	2
(98)	9 + 0	9
(99)	1 + 4	5
(100)	2 + 1	3

Addition 2

Name______________________________ Date ________________ Score ______ Time______

(1) 2+7
(2) 9+2
(3) 3+2
(4) 6+2
(5) 2+2
(6) 5+2
(7) 4+2
(8) 2+10
(9) 2+6
(10) 2+2
(11) 2+4
(12) 2+6
(13) 2+7
(14) 9+2
(15) 5+2
(16) 4+2
(17) 2+1
(18) 2+8
(19) 1+2
(20) 2+0
(21) 8+2
(22) 2+7
(23) 2+9
(24) 3+2
(25) 2+4
(26) 2+0
(27) 1+2
(28) 7+2
(29) 2+6
(30) 2+7
(31) 2+4
(32) 2+1
(33) 2+9
(34) 4+2
(35) 2+7
(36) 2+2
(37) 5+2
(38) 3+2
(39) 0+2
(40) 1+2
(41) 2+7
(42) 2+9
(43) 7+2
(44) 2+8
(45) 2+10
(46) 2+5
(47) 1+2
(48) 2+4
(49) 2+1
(50) 0+2
(51) 2+9
(52) 4+2
(53) 2+7
(54) 2+2
(55) 5+2
(56) 3+2
(57) 0+2
(58) 1+2
(59) 2+1
(60) 9+2
(61) 2+4
(62) 4+2
(63) 6+2
(64) 2+5
(65) 2+2
(66) 2+6
(67) 2+8
(68) 3+2
(69) 10+2
(70) 7+2
(71) 2+0
(72) 5+2
(73) 2+10
(74) 8+2
(75) 2+7
(76) 1+2
(77) 2+3
(78) 2+9
(79) 0+2
(80) 10+2
(81) 7+2
(82) 3+2
(83) 2+0
(84) 2+8
(85) 5+2
(86) 2+6
(87) 2+10
(88) 2+2
(89) 8+2
(90) 2+5
(91) 2+7
(92) 6+2
(93) 1+2
(94) 4+2
(95) 2+3
(96) 2+4
(97) 2+9
(98) 9+2
(99) 0+2
(100) 2+1

Addition 2

KEY

(1)	2 + 7	9
(2)	9 + 2	11
(3)	3 + 2	5
(4)	6 + 2	8
(5)	2 + 2	4
(6)	5 + 2	7
(7)	4 + 2	6
(8)	2 + 10	12
(9)	2 + 6	8
(10)	2 + 2	4
(11)	2 + 4	6
(12)	2 + 6	8
(13)	2 + 7	9
(14)	9 + 2	11
(15)	5 + 2	7
(16)	4 + 2	6
(17)	2 + 1	3
(18)	2 + 8	10
(19)	1 + 2	3
(20)	2 + 0	2
(21)	8 + 2	10
(22)	2 + 7	9
(23)	2 + 9	11
(24)	3 + 2	5
(25)	2 + 4	6
(26)	2 + 0	2
(27)	1 + 2	3
(28)	7 + 2	9
(29)	2 + 6	8
(30)	2 + 7	9
(31)	2 + 4	6
(32)	2 + 1	3
(33)	2 + 9	11
(34)	4 + 2	6
(35)	2 + 7	9
(36)	2 + 2	4
(37)	5 + 2	7
(38)	3 + 2	5
(39)	0 + 2	2
(40)	1 + 2	3
(41)	2 + 7	9
(42)	2 + 9	11
(43)	7 + 2	9
(44)	2 + 8	10
(45)	2 + 10	12
(46)	2 + 5	7
(47)	1 + 2	3
(48)	2 + 4	6
(49)	2 + 1	3
(50)	0 + 2	2
(51)	2 + 9	11
(52)	4 + 2	6
(53)	2 + 7	9
(54)	2 + 2	4
(55)	5 + 2	7
(56)	3 + 2	5
(57)	0 + 2	2
(58)	1 + 2	3
(59)	2 + 1	3
(60)	9 + 2	11
(61)	2 + 4	6
(62)	4 + 2	6
(63)	6 + 2	8
(64)	2 + 5	7
(65)	2 + 2	4
(66)	2 + 6	8
(67)	2 + 8	10
(68)	3 + 2	5
(69)	10 + 2	12
(70)	7 + 2	9
(71)	2 + 0	2
(72)	5 + 2	7
(73)	2 + 10	12
(74)	8 + 2	10
(75)	2 + 7	9
(76)	1 + 2	3
(77)	2 + 3	5
(78)	2 + 9	11
(79)	0 + 2	2
(80)	10 + 2	12
(81)	7 + 2	9
(82)	3 + 2	5
(83)	2 + 0	2
(84)	2 + 8	10
(85)	5 + 2	7
(86)	2 + 6	8
(87)	2 + 10	12
(88)	2 + 2	4
(89)	8 + 2	10
(90)	2 + 5	7
(91)	2 + 7	9
(92)	6 + 2	8
(93)	1 + 2	3
(94)	4 + 2	6
(95)	2 + 3	5
(96)	2 + 4	6
(97)	2 + 9	11
(98)	9 + 2	11
(99)	0 + 2	2
(100)	2 + 1	3

Addition 3

Name ____________________ Date ______________ Score ______ Time ______

(1) $6 + 3$
(2) $3 + 5$
(3) $8 + 3$
(4) $3 + 0$
(5) $9 + 3$
(6) $3 + 3$
(7) $7 + 3$
(8) $3 + 5$
(9) $6 + 3$
(10) $3 + 8$
(11) $1 + 3$
(12) $3 + 10$
(13) $2 + 3$
(14) $8 + 3$
(15) $3 + 2$
(16) $3 + 7$
(17) $3 + 9$
(18) $10 + 3$
(19) $5 + 3$
(20) $3 + 6$
(21) $3 + 3$
(22) $0 + 3$
(23) $3 + 1$
(24) $4 + 3$
(25) $3 + 4$
(26) $9 + 3$
(27) $3 + 7$
(28) $3 + 4$
(29) $0 + 3$
(30) $3 + 8$
(31) $4 + 3$
(32) $3 + 1$
(33) $3 + 8$
(34) $0 + 3$
(35) $3 + 3$
(36) $3 + 5$
(37) $3 + 6$
(38) $6 + 3$
(39) $5 + 3$
(40) $9 + 3$
(41) $10 + 3$
(42) $3 + 8$
(43) $3 + 9$
(44) $8 + 3$
(45) $3 + 7$
(46) $10 + 3$
(47) $3 + 2$
(48) $0 + 3$
(49) $1 + 3$
(50) $3 + 5$
(51) $3 + 0$
(52) $3 + 5$
(53) $3 + 10$
(54) $3 + 7$
(55) $3 + 6$
(56) $3 + 1$
(57) $4 + 3$
(58) $5 + 3$
(59) $3 + 2$
(60) $1 + 3$
(61) $7 + 3$
(62) $3 + 8$
(63) $3 + 4$
(64) $0 + 3$
(65) $10 + 3$
(66) $8 + 3$
(67) $3 + 9$
(68) $3 + 8$
(69) $6 + 3$
(70) $3 + 5$
(71) $3 + 3$
(72) $3 + 8$
(73) $10 + 3$
(74) $0 + 3$
(75) $8 + 3$
(76) $2 + 3$
(77) $3 + 4$
(78) $8 + 3$
(79) $3 + 5$
(80) $3 + 10$
(81) $7 + 3$
(82) $1 + 3$
(83) $3 + 8$
(84) $3 + 2$
(85) $6 + 3$
(86) $5 + 3$
(87) $3 + 1$
(88) $3 + 5$
(89) $7 + 3$
(90) $4 + 3$
(91) $3 + 3$
(92) $3 + 6$
(93) $9 + 3$
(94) $3 + 7$
(95) $3 + 0$
(96) $3 + 10$
(97) $8 + 3$
(98) $3 + 5$
(99) $3 + 0$
(100) $6 + 3$

Addition 3

KEY

(1)	$6+3$	9	(26)	$9+3$	12	(51)	$3+0$	3	(76)	$2+3$	5
(2)	$3+5$	8	(27)	$3+7$	10	(52)	$3+5$	8	(77)	$3+4$	7
(3)	$8+3$	11	(28)	$3+4$	7	(53)	$3+10$	13	(78)	$8+3$	11
(4)	$3+0$	3	(29)	$0+3$	3	(54)	$3+7$	10	(79)	$3+5$	8
(5)	$9+3$	12	(30)	$3+8$	11	(55)	$3+6$	9	(80)	$3+10$	13
(6)	$3+3$	6	(31)	$4+3$	7	(56)	$3+1$	4	(81)	$7+3$	10
(7)	$7+3$	10	(32)	$3+1$	4	(57)	$4+3$	7	(82)	$1+3$	4
(8)	$3+5$	8	(33)	$3+8$	11	(58)	$5+3$	8	(83)	$3+8$	11
(9)	$6+3$	9	(34)	$0+3$	3	(59)	$3+2$	5	(84)	$3+2$	5
(10)	$3+8$	11	(35)	$3+3$	6	(60)	$1+3$	4	(85)	$6+3$	9
(11)	$1+3$	4	(36)	$3+5$	8	(61)	$7+3$	10	(86)	$5+3$	8
(12)	$3+10$	13	(37)	$3+6$	9	(62)	$3+8$	11	(87)	$3+1$	4
(13)	$2+3$	5	(38)	$6+3$	9	(63)	$3+4$	7	(88)	$3+5$	8
(14)	$8+3$	11	(39)	$5+3$	8	(64)	$0+3$	3	(89)	$7+3$	10
(15)	$3+2$	5	(40)	$9+3$	12	(65)	$10+3$	13	(90)	$4+3$	7
(16)	$3+7$	10	(41)	$10+3$	13	(66)	$8+3$	11	(91)	$3+3$	6
(17)	$3+9$	12	(42)	$3+8$	11	(67)	$3+9$	12	(92)	$3+6$	9
(18)	$10+3$	13	(43)	$3+9$	12	(68)	$3+8$	11	(93)	$9+3$	12
(19)	$5+3$	8	(44)	$8+3$	11	(69)	$6+3$	9	(94)	$3+7$	10
(20)	$3+6$	9	(45)	$3+7$	10	(70)	$3+5$	8	(95)	$3+0$	3
(21)	$3+3$	6	(46)	$10+3$	13	(71)	$3+3$	6	(96)	$3+10$	13
(22)	$0+3$	3	(47)	$3+2$	5	(72)	$3+8$	11	(97)	$8+3$	11
(23)	$3+1$	4	(48)	$0+3$	3	(73)	$10+3$	13	(98)	$3+5$	8
(24)	$4+3$	7	(49)	$1+3$	4	(74)	$0+3$	3	(99)	$3+0$	3
(25)	$3+4$	7	(50)	$3+5$	8	(75)	$8+3$	11	(100)	$6+3$	9

Addition 0-3

Name______________________________ Date _______________ Score ______ Time______

(1)	2 + 6		(26)	1 + 1		(51)	2 + 5		(76)	2 + 8	
(2)	1 + 3		(27)	2 + 10		(52)	3 + 4		(77)	3 + 5	
(3)	3 + 5		(28)	1 + 8		(53)	2 + 8		(78)	3 + 8	
(4)	3 + 9		(29)	1 + 3		(54)	1 + 8		(79)	1 + 10	
(5)	2 + 4		(30)	3 + 5		(55)	2 + 3		(80)	2 + 9	
(6)	2 + 7		(31)	1 + 8		(56)	3 + 10		(81)	3 + 6	
(7)	3 + 7		(32)	2 + 1		(57)	1 + 6		(82)	2 + 3	
(8)	1 + 6		(33)	3 + 8		(58)	2 + 9		(83)	2 + 7	
(9)	1 + 8		(34)	3 + 4		(59)	1 + 4		(84)	1 + 7	
(10)	2 + 5		(35)	3 + 10		(60)	3 + 7		(85)	1 + 1	
(11)	2 + 1		(36)	8 + 1		(61)	1 + 10		(86)	1 + 3	
(12)	2 + 6		(37)	2 + 9		(62)	1 + 6		(87)	2 + 1	
(13)	3 + 6		(38)	3 + 7		(63)	2 + 7		(88)	1 + 8	
(14)	1 + 5		(39)	1 + 6		(64)	3 + 2		(89)	3 + 7	
(15)	2 + 5		(40)	3 + 2		(65)	1 + 7		(90)	2 + 4	
(16)	2 + 8		(41)	2 + 4		(66)	7 + 3		(91)	3 + 5	
(17)	2 + 3		(42)	2 + 2		(67)	2 + 4		(92)	3 + 8	
(18)	1 + 6		(43)	1 + 1		(68)	2 + 2		(93)	3 + 6	
(19)	1 + 4		(44)	3 + 5		(69)	3 + 9		(94)	2 + 8	
(20)	2 + 7		(45)	2 + 6		(70)	1 + 1		(95)	3 + 10	
(21)	1 + 10		(46)	1 + 8		(71)	2 + 10		(96)	1 + 4	
(22)	3 + 7		(47)	3 + 6		(72)	3 + 5		(97)	1 + 6	
(23)	1 + 7		(48)	2 + 1		(73)	1 + 8		(98)	2 + 2	
(24)	2 + 10		(49)	1 + 5		(74)	2 + 6		(99)	1 + 7	
(25)	2 + 2		(50)	3 + 8		(75)	1 + 3		(100)	2 + 10	

Addition 0-3

KEY

#	Problem	Answer	#	Problem	Answer	#	Problem	Answer	#	Problem	Answer
(1)	2 + 6	8	(26)	1 + 1	2	(51)	2 + 5	7	(76)	2 + 8	10
(2)	1 + 3	4	(27)	2 + 10	12	(52)	3 + 4	7	(77)	3 + 5	8
(3)	3 + 5	8	(28)	1 + 8	9	(53)	2 + 8	10	(78)	3 + 8	11
(4)	3 + 9	12	(29)	1 + 3	4	(54)	1 + 8	9	(79)	1 + 10	11
(5)	2 + 4	6	(30)	3 + 5	8	(55)	2 + 3	5	(80)	2 + 9	11
(6)	2 + 7	9	(31)	1 + 8	9	(56)	3 + 10	13	(81)	3 + 6	9
(7)	3 + 7	10	(32)	2 + 1	3	(57)	1 + 6	7	(82)	2 + 3	5
(8)	1 + 6	7	(33)	3 + 8	11	(58)	2 + 9	11	(83)	2 + 7	9
(9)	1 + 8	9	(34)	3 + 4	7	(59)	1 + 4	5	(84)	1 + 7	8
(10)	2 + 5	7	(35)	3 + 10	13	(60)	3 + 7	10	(85)	1 + 1	2
(11)	2 + 1	3	(36)	8 + 1	9	(61)	1 + 10	11	(86)	1 + 3	4
(12)	2 + 6	8	(37)	2 + 9	11	(62)	1 + 6	7	(87)	2 + 1	3
(13)	3 + 6	9	(38)	3 + 7	10	(63)	2 + 7	9	(88)	1 + 8	9
(14)	1 + 5	6	(39)	1 + 6	7	(64)	3 + 2	5	(89)	3 + 7	10
(15)	2 + 5	7	(40)	3 + 2	5	(65)	1 + 7	8	(90)	2 + 4	6
(16)	2 + 8	10	(41)	2 + 4	6	(66)	7 + 3	10	(91)	3 + 5	8
(17)	2 + 3	5	(42)	2 + 2	4	(67)	2 + 4	6	(92)	3 + 8	11
(18)	1 + 6	7	(43)	1 + 1	2	(68)	2 + 2	4	(93)	3 + 6	9
(19)	1 + 4	5	(44)	3 + 5	8	(69)	3 + 9	12	(94)	2 + 8	10
(20)	2 + 7	9	(45)	2 + 6	8	(70)	1 + 1	2	(95)	3 + 10	13
(21)	1 + 10	11	(46)	1 + 8	9	(71)	2 + 10	12	(96)	1 + 4	5
(22)	3 + 7	10	(47)	3 + 6	9	(72)	3 + 5	8	(97)	1 + 6	7
(23)	1 + 7	8	(48)	2 + 1	3	(73)	1 + 8	9	(98)	2 + 2	4
(24)	2 + 10	12	(49)	1 + 5	6	(74)	2 + 6	8	(99)	1 + 7	8
(25)	2 + 2	4	(50)	3 + 8	11	(75)	1 + 3	4	(100)	2 + 10	12

Addition 4

Name____________________ Date __________ Score _____ Time_____

(1) 4 + 3
(2) 4 + 9
(3) 2 + 4
(4) 4 + 8
(5) 6 + 4
(6) 4 + 0
(7) 4 + 7
(8) 4 + 3
(9) 4 + 6
(10) 4 + 4
(11) 9 + 4
(12) 3 + 4
(13) 4 + 8
(14) 4 + 5
(15) 7 + 4
(16) 4 + 1
(17) 5 + 4
(18) 4 + 7
(19) 4 + 4
(20) 4 + 2
(21) 7 + 4
(22) 0 + 4
(23) 3 + 4
(24) 1 + 4
(25) 4 + 10
(26) 2 + 4
(27) 0 + 4
(28) 4 + 7
(29) 4 + 5
(30) 4 + 10
(31) 4 + 0
(32) 8 + 4
(33) 4 + 9
(34) 4 + 2
(35) 4 + 5
(36) 7 + 4
(37) 9 + 4
(38) 4 + 4
(39) 4 + 1
(40) 4 + 8
(41) 5 + 4
(42) 4 + 7
(43) 0 + 4
(44) 3 + 4
(45) 4 + 9
(46) 4 + 6
(47) 10 + 4
(48) 4 + 3
(49) 8 + 4
(50) 2 + 4
(51) 4 + 1
(52) 7 + 4
(53) 3 + 4
(54) 4 + 8
(55) 4 + 0
(56) 4 + 10
(57) 1 + 4
(58) 7 + 4
(59) 4 + 4
(60) 4 + 1
(61) 4 + 8
(62) 3 + 4
(63) 4 + 6
(64) 4 + 3
(65) 6 + 4
(66) 10 + 4
(67) 2 + 4
(68) 4 + 9
(69) 0 + 4
(70) 5 + 4
(71) 4 + 7
(72) 9 + 4
(73) 4 + 5
(74) 4 + 10
(75) 4 + 0
(76) 6 + 4
(77) 10 + 4
(78) 4 + 3
(79) 2 + 4
(80) 4 + 6
(81) 4 + 9
(82) 3 + 4
(83) 0 + 4
(84) 4 + 8
(85) 5 + 4
(86) 4 + 1
(87) 4 + 7
(88) 4 + 4
(89) 4 + 9
(90) 7 + 4
(91) 4 + 5
(92) 1 + 4
(93) 4 + 2
(94) 4 + 10
(95) 8 + 4
(96) 0 + 4
(97) 4 + 4
(98) 5 + 4
(99) 9 + 4
(100) 4 + 2

Addition 4

KEY

#	Problem	Answer
(1)	4 + 3	7
(2)	4 + 9	13
(3)	2 + 4	6
(4)	4 + 8	12
(5)	6 + 4	10
(6)	4 + 0	4
(7)	4 + 7	11
(8)	4 + 3	7
(9)	4 + 6	10
(10)	4 + 4	8
(11)	9 + 4	13
(12)	3 + 4	7
(13)	4 + 8	12
(14)	4 + 5	9
(15)	7 + 4	11
(16)	4 + 1	5
(17)	5 + 4	9
(18)	4 + 7	11
(19)	4 + 4	8
(20)	4 + 2	6
(21)	7 + 4	11
(22)	0 + 4	4
(23)	3 + 4	7
(24)	1 + 4	5
(25)	4 + 10	14
(26)	2 + 4	6
(27)	0 + 4	4
(28)	4 + 7	11
(29)	4 + 5	9
(30)	4 + 10	14
(31)	4 + 0	4
(32)	8 + 4	12
(33)	4 + 9	13
(34)	4 + 2	6
(35)	4 + 5	9
(36)	7 + 4	11
(37)	9 + 4	13
(38)	4 + 4	8
(39)	4 + 1	5
(40)	4 + 8	12
(41)	5 + 4	9
(42)	4 + 7	11
(43)	0 + 4	4
(44)	3 + 4	7
(45)	4 + 9	13
(46)	4 + 6	10
(47)	10 + 4	14
(48)	4 + 3	7
(49)	8 + 4	12
(50)	2 + 4	6
(51)	4 + 1	5
(52)	7 + 4	11
(53)	3 + 4	7
(54)	4 + 8	12
(55)	4 + 0	4
(56)	4 + 10	14
(57)	1 + 4	5
(58)	7 + 4	11
(59)	4 + 4	8
(60)	4 + 1	5
(61)	4 + 8	12
(62)	3 + 4	7
(63)	4 + 6	10
(64)	4 + 3	7
(65)	6 + 4	10
(66)	10 + 4	14
(67)	2 + 4	6
(68)	4 + 9	13
(69)	0 + 4	4
(70)	5 + 4	9
(71)	4 + 7	11
(72)	9 + 4	13
(73)	4 + 5	9
(74)	4 + 10	14
(75)	4 + 0	4
(76)	6 + 4	10
(77)	10 + 4	14
(78)	4 + 3	7
(79)	2 + 4	6
(80)	4 + 6	10
(81)	4 + 9	13
(82)	3 + 4	7
(83)	0 + 4	4
(84)	4 + 8	12
(85)	5 + 4	9
(86)	4 + 1	5
(87)	4 + 7	11
(88)	4 + 4	8
(89)	4 + 9	13
(90)	7 + 4	11
(91)	4 + 5	9
(92)	1 + 4	5
(93)	4 + 2	6
(94)	4 + 10	14
(95)	8 + 4	12
(96)	0 + 4	4
(97)	4 + 4	8
(98)	5 + 4	9
(99)	9 + 4	13
(100)	4 + 2	6

Addition 0-4

Name ______________________________ Date ______________ Score ______ Time ______

(1) $6 + 4$ ☐
(2) $1 + 4$ ☐
(3) $2 + 2$ ☐
(4) $4 + 8$ ☐
(5) $2 + 1$ ☐
(6) $3 + 3$ ☐
(7) $4 + 7$ ☐
(8) $1 + 8$ ☐
(9) $4 + 2$ ☐
(10) $1 + 3$ ☐
(11) $3 + 7$ ☐
(12) $4 + 4$ ☐
(13) $2 + 5$ ☐
(14) $4 + 10$ ☐
(15) $1 + 2$ ☐
(16) $4 + 5$ ☐
(17) $2 + 9$ ☐
(18) $4 + 1$ ☐
(19) $3 + 4$ ☐
(20) $4 + 6$ ☐
(21) $3 + 10$ ☐
(22) $4 + 3$ ☐
(23) $7 + 1$ ☐
(24) $4 + 9$ ☐
(25) $2 + 4$ ☐

(26) $3 + 7$ ☐
(27) $3 + 10$ ☐
(28) $1 + 7$ ☐
(29) $2 + 10$ ☐
(30) $2 + 3$ ☐
(31) $3 + 5$ ☐
(32) $4 + 6$ ☐
(33) $1 + 4$ ☐
(34) $4 + 8$ ☐
(35) $2 + 8$ ☐
(36) $3 + 3$ ☐
(37) $4 + 1$ ☐
(38) $1 + 6$ ☐
(39) $4 + 7$ ☐
(40) $2 + 5$ ☐
(41) $1 + 8$ ☐
(42) $4 + 2$ ☐
(43) $3 + 6$ ☐
(44) $4 + 10$ ☐
(45) $1 + 8$ ☐
(46) $2 + 2$ ☐
(47) $1 + 5$ ☐
(48) $4 + 9$ ☐
(49) $4 + 2$ ☐
(50) $3 + 4$ ☐

(51) $1 + 10$ ☐
(52) $4 + 4$ ☐
(53) $2 + 6$ ☐
(54) $4 + 9$ ☐
(55) $3 + 2$ ☐
(56) $4 + 1$ ☐
(57) $1 + 1$ ☐
(58) $3 + 8$ ☐
(59) $1 + 9$ ☐
(60) $4 + 10$ ☐
(61) $3 + 4$ ☐
(62) $4 + 8$ ☐
(63) $1 + 2$ ☐
(64) $4 + 2$ ☐
(65) $2 + 7$ ☐
(66) $3 + 1$ ☐
(67) $4 + 5$ ☐
(68) $2 + 9$ ☐
(69) $1 + 10$ ☐
(70) $3 + 9$ ☐
(71) $4 + 3$ ☐
(72) $3 + 6$ ☐
(73) $4 + 7$ ☐
(74) $2 + 3$ ☐
(75) $2 + 1$ ☐

(76) $4 + 10$ ☐
(77) $3 + 8$ ☐
(78) $6 + 2$ ☐
(79) $4 + 2$ ☐
(80) $2 + 7$ ☐
(81) $1 + 5$ ☐
(82) $3 + 2$ ☐
(83) $4 + 5$ ☐
(84) $4 + 9$ ☐
(85) $3 + 5$ ☐
(86) $2 + 10$ ☐
(87) $1 + 9$ ☐
(88) $3 + 6$ ☐
(89) $4 + 1$ ☐
(90) $2 + 8$ ☐
(91) $4 + 3$ ☐
(92) $3 + 9$ ☐
(93) $4 + 8$ ☐
(94) $2 + 3$ ☐
(95) $1 + 1$ ☐
(96) $4 + 7$ ☐
(97) $3 + 1$ ☐
(98) $4 + 4$ ☐
(99) $1 + 6$ ☐
(100) $4 + 6$ ☐

Addition 0-4

KEY

No.	Problem	Answer	No.	Problem	Answer	No.	Problem	Answer	No.	Problem	Answer
(1)	$6+4$	10	(26)	$3+7$	10	(51)	$1+10$	11	(76)	$4+10$	14
(2)	$1+4$	5	(27)	$3+10$	13	(52)	$4+4$	8	(77)	$3+8$	11
(3)	$2+2$	4	(28)	$1+7$	8	(53)	$2+6$	8	(78)	$6+2$	8
(4)	$4+8$	12	(29)	$2+10$	12	(54)	$4+9$	13	(79)	$4+2$	6
(5)	$2+1$	3	(30)	$2+3$	5	(55)	$3+2$	5	(80)	$2+7$	9
(6)	$3+3$	6	(31)	$3+5$	8	(56)	$4+1$	5	(81)	$1+5$	6
(7)	$4+7$	11	(32)	$4+6$	10	(57)	$1+1$	2	(82)	$3+2$	5
(8)	$1+8$	9	(33)	$1+4$	5	(58)	$3+8$	11	(83)	$4+5$	9
(9)	$4+2$	6	(34)	$4+8$	12	(59)	$1+9$	10	(84)	$4+9$	13
(10)	$1+3$	4	(35)	$2+8$	10	(60)	$4+10$	14	(85)	$3+5$	8
(11)	$3+7$	10	(36)	$3+3$	6	(61)	$3+4$	7	(86)	$2+10$	12
(12)	$4+4$	8	(37)	$4+1$	5	(62)	$4+8$	12	(87)	$1+9$	10
(13)	$2+5$	7	(38)	$1+6$	7	(63)	$1+2$	3	(88)	$3+6$	9
(14)	$4+10$	14	(39)	$4+7$	11	(64)	$4+2$	6	(89)	$4+1$	5
(15)	$1+2$	3	(40)	$2+5$	7	(65)	$2+7$	9	(90)	$2+8$	10
(16)	$4+5$	9	(41)	$1+8$	9	(66)	$3+1$	4	(91)	$4+3$	7
(17)	$2+9$	11	(42)	$4+2$	6	(67)	$4+5$	9	(92)	$3+9$	12
(18)	$4+1$	5	(43)	$3+6$	9	(68)	$2+9$	11	(93)	$4+8$	12
(19)	$3+4$	7	(44)	$4+10$	14	(69)	$1+10$	11	(94)	$2+3$	5
(20)	$4+6$	10	(45)	$1+8$	9	(70)	$3+9$	12	(95)	$1+1$	2
(21)	$3+10$	13	(46)	$2+2$	4	(71)	$4+3$	7	(96)	$4+7$	11
(22)	$4+3$	7	(47)	$1+5$	6	(72)	$3+6$	9	(97)	$3+1$	4
(23)	$7+1$	8	(48)	$4+9$	13	(73)	$4+7$	11	(98)	$4+4$	8
(24)	$4+9$	13	(49)	$4+2$	6	(74)	$2+3$	5	(99)	$1+6$	7
(25)	$2+4$	6	(50)	$3+4$	7	(75)	$2+1$	3	(100)	$4+6$	10

Addition 5

Name ______________________________ Date ________________ Score ______ Time ______

(1) 5 + 3
(2) 10 + 5
(3) 5 + 4
(4) 5 + 2
(5) 8 + 5
(6) 5 + 9
(7) 2 + 5
(8) 5 + 5
(9) 5 + 0
(10) 5 + 6
(11) 5 + 10
(12) 1 + 5
(13) 3 + 5
(14) 7 + 5
(15) 6 + 5
(16) 5 + 1
(17) 5 + 8
(18) 9 + 5
(19) 0 + 5
(20) 4 + 5
(21) 5 + 7
(22) 5 + 5
(23) 5 + 2
(24) 8 + 5
(25) 5 + 10
(26) 5 + 7
(27) 0 + 5
(28) 5 + 8
(29) 6 + 5
(30) 3 + 5
(31) 5 + 10
(32) 5 + 0
(33) 2 + 5
(34) 8 + 5
(35) 5 + 4
(36) 5 + 3
(37) 10 + 5
(38) 5 + 2
(39) 5 + 9
(40) 5 + 5
(41) 5 + 6
(42) 1 + 5
(43) 7 + 5
(44) 5 + 1
(45) 9 + 5
(46) 4 + 5
(47) 7 + 5
(48) 5 + 8
(49) 6 + 5
(50) 5 + 3
(51) 9 + 5
(52) 5 + 0
(53) 5 + 7
(54) 10 + 5
(55) 6 + 5
(56) 5 + 3
(57) 7 + 5
(58) 5 + 1
(59) 10 + 5
(60) 3 + 5
(61) 5 + 8
(62) 5 + 4
(63) 1 + 5
(64) 9 + 5
(65) 5 + 2
(66) 5 + 10
(67) 0 + 5
(68) 8 + 5
(69) 5 + 6
(70) 4 + 5
(71) 5 + 9
(72) 5 + 0
(73) 5 + 7
(74) 2 + 5
(75) 5 + 5
(76) 5 + 6
(77) 5 + 9
(78) 2 + 5
(79) 4 + 5
(80) 5 + 3
(81) 5 + 1
(82) 3 + 5
(83) 5 + 4
(84) 9 + 5
(85) 5 + 10
(86) 8 + 5
(87) 4 + 5
(88) 5 + 0
(89) 2 + 5
(90) 5 + 5
(91) 5 + 7
(92) 5 + 9
(93) 5 + 6
(94) 0 + 5
(95) 5 + 2
(96) 1 + 5
(97) 5 + 8
(98) 10 + 5
(99) 7 + 5
(100) 6 + 5

Addition 5

KEY

No.	Problem	Answer
(1)	$5 + 3$	8
(2)	$10 + 5$	15
(3)	$5 + 4$	9
(4)	$5 + 2$	7
(5)	$8 + 5$	13
(6)	$5 + 9$	14
(7)	$2 + 5$	7
(8)	$5 + 5$	10
(9)	$5 + 0$	5
(10)	$5 + 6$	11
(11)	$5 + 10$	15
(12)	$1 + 5$	6
(13)	$3 + 5$	8
(14)	$7 + 5$	12
(15)	$6 + 5$	11
(16)	$5 + 1$	6
(17)	$5 + 8$	13
(18)	$9 + 5$	14
(19)	$0 + 5$	5
(20)	$4 + 5$	9
(21)	$5 + 7$	12
(22)	$5 + 5$	10
(23)	$5 + 2$	7
(24)	$8 + 5$	13
(25)	$5 + 10$	15
(26)	$5 + 7$	12
(27)	$0 + 5$	5
(28)	$5 + 8$	13
(29)	$6 + 5$	11
(30)	$3 + 5$	8
(31)	$5 + 10$	15
(32)	$5 + 0$	5
(33)	$2 + 5$	7
(34)	$8 + 5$	13
(35)	$5 + 4$	9
(36)	$5 + 3$	8
(37)	$10 + 5$	15
(38)	$5 + 2$	7
(39)	$5 + 9$	14
(40)	$5 + 5$	10
(41)	$5 + 6$	11
(42)	$1 + 5$	6
(43)	$7 + 5$	12
(44)	$5 + 1$	6
(45)	$9 + 5$	14
(46)	$4 + 5$	9
(47)	$7 + 5$	12
(48)	$5 + 8$	13
(49)	$6 + 5$	11
(50)	$5 + 3$	8
(51)	$9 + 5$	14
(52)	$5 + 0$	5
(53)	$5 + 7$	12
(54)	$10 + 5$	15
(55)	$6 + 5$	11
(56)	$5 + 3$	8
(57)	$7 + 5$	12
(58)	$5 + 1$	6
(59)	$10 + 5$	15
(60)	$3 + 5$	8
(61)	$5 + 8$	13
(62)	$5 + 4$	9
(63)	$1 + 5$	6
(64)	$9 + 5$	14
(65)	$5 + 2$	7
(66)	$5 + 10$	15
(67)	$0 + 5$	5
(68)	$8 + 5$	13
(69)	$5 + 6$	11
(70)	$4 + 5$	9
(71)	$5 + 9$	14
(72)	$5 + 0$	5
(73)	$5 + 7$	12
(74)	$2 + 5$	7
(75)	$5 + 5$	10
(76)	$5 + 6$	11
(77)	$5 + 9$	14
(78)	$2 + 5$	7
(79)	$4 + 5$	9
(80)	$5 + 3$	8
(81)	$5 + 1$	6
(82)	$3 + 5$	8
(83)	$5 + 4$	9
(84)	$9 + 5$	14
(85)	$5 + 10$	15
(86)	$8 + 5$	13
(87)	$4 + 5$	9
(88)	$5 + 0$	5
(89)	$2 + 5$	7
(90)	$5 + 5$	10
(91)	$5 + 7$	12
(92)	$5 + 9$	14
(93)	$5 + 6$	11
(94)	$0 + 5$	5
(95)	$5 + 2$	7
(96)	$1 + 5$	6
(97)	$5 + 8$	13
(98)	$10 + 5$	15
(99)	$7 + 5$	12
(100)	$6 + 5$	11

Addition 0-5

Name ______________________________ Date ______________ Score ______ Time ______

(1) 3 + 8
(2) 0 + 9
(3) 5 + 10
(4) 2 + 7
(5) 1 + 9
(6) 4 + 6
(7) 5 + 3
(8) 4 + 10
(9) 5 + 5
(10) 2 + 6
(11) 0 + 7
(12) 3 + 2
(13) 4 + 2
(14) 5 + 7
(15) 1 + 3
(16) 2 + 9
(17) 3 + 6
(18) 5 + 2
(19) 0 + 6
(20) 5 + 9
(21) 2 + 3
(22) 3 + 10
(23) 4 + 4
(24) 5 + 1
(25) 0 + 0
(26) 5 + 0
(27) 2 + 8
(28) 5 + 7
(29) 1 + 6
(30) 4 + 0
(31) 2 + 4
(32) 4 + 8
(33) 5 + 3
(34) 2 + 0
(35) 4 + 3
(36) 5 + 5
(37) 0 + 8
(38) 5 + 4
(39) 3 + 5
(40) 2 + 2
(41) 0 + 3
(42) 5 + 8
(43) 4 + 9
(44) 3 + 1
(45) 5 + 6
(46) 4 + 7
(47) 2 + 10
(48) 0 + 5
(49) 1 + 0
(50) 3 + 9
(51) 2 + 1
(52) 3 + 8
(53) 5 + 8
(54) 3 + 0
(55) 4 + 7
(56) 2 + 8
(57) 5 + 0
(58) 4 + 9
(59) 0 + 1
(60) 3 + 3
(61) 4 + 5
(62) 5 + 2
(63) 3 + 9
(64) 1 + 7
(65) 5 + 10
(66) 3 + 7
(67) 5 + 1
(68) 5 + 6
(69) 3 + 4
(70) 2 + 5
(71) 5 + 9
(72) 0 + 2
(73) 4 + 1
(74) 5 + 4
(75) 0 + 10
(76) 4 + 8
(77) 5 + 10
(78) 3 + 5
(79) 5 + 2
(80) 4 + 6
(81) 5 + 7
(82) 0 + 6
(83) 1 + 1
(84) 5 + 5
(85) 4 + 10
(86) 5 + 8
(87) 3 + 3
(88) 5 + 4
(89) 0 + 7
(90) 1 + 5
(91) 3 + 10
(92) 4 + 5
(93) 5 + 0
(94) 3 + 7
(95) 5 + 6
(96) 4 + 4
(97) 3 + 6
(98) 5 + 9
(99) 0 + 4
(100) 4 + 8

Addition 0-5

KEY

(1)	3+8	11	(26)	5+0	5	(51)	2+1	3	(76)	4+8	12
(2)	0+9	9	(27)	2+8	10	(52)	3+8	11	(77)	5+10	15
(3)	5+10	15	(28)	5+7	12	(53)	5+8	13	(78)	3+5	8
(4)	2+7	9	(29)	1+6	7	(54)	3+0	3	(79)	5+2	7
(5)	1+9	10	(30)	4+0	4	(55)	4+7	11	(80)	4+6	10
(6)	4+6	10	(31)	2+4	6	(56)	2+8	10	(81)	5+7	12
(7)	5+3	8	(32)	4+8	12	(57)	5+0	5	(82)	0+6	6
(8)	4+10	14	(33)	5+3	8	(58)	4+9	13	(83)	1+1	2
(9)	5+5	10	(34)	2+0	2	(59)	0+1	1	(84)	5+5	10
(10)	2+6	8	(35)	4+3	7	(60)	3+3	6	(85)	4+10	14
(11)	0+7	7	(36)	5+5	10	(61)	4+5	9	(86)	5+8	13
(12)	3+2	5	(37)	0+8	8	(62)	5+2	7	(87)	3+3	6
(13)	4+2	6	(38)	5+4	9	(63)	3+9	12	(88)	5+4	9
(14)	5+7	12	(39)	3+5	8	(64)	1+7	8	(89)	0+7	7
(15)	1+3	4	(40)	2+2	4	(65)	5+10	15	(90)	1+5	6
(16)	2+9	11	(41)	0+3	3	(66)	3+7	10	(91)	3+10	13
(17)	3+6	9	(42)	5+8	13	(67)	5+1	6	(92)	4+5	9
(18)	5+2	7	(43)	4+9	13	(68)	5+6	11	(93)	5+0	5
(19)	0+6	6	(44)	3+1	4	(69)	3+4	7	(94)	3+7	10
(20)	5+9	14	(45)	5+6	11	(70)	2+5	7	(95)	5+6	11
(21)	2+3	5	(46)	4+7	11	(71)	5+9	14	(96)	4+4	8
(22)	3+10	13	(47)	2+10	12	(72)	0+2	2	(97)	3+6	9
(23)	4+4	8	(48)	0+5	5	(73)	4+1	5	(98)	5+9	14
(24)	5+1	6	(49)	1+0	1	(74)	5+4	9	(99)	0+4	4
(25)	0+0	0	(50)	3+9	12	(75)	0+10	10	(100)	4+8	12

Addition 6

Name ______________________________ Date _______________ Score ______ Time ______

(1) 4 + 6
(2) 6 + 7
(3) 2 + 6
(4) 6 + 5
(5) 1 + 6
(6) 6 + 9
(7) 6 + 2
(8) 9 + 6
(9) 1 + 6
(10) 6 + 10
(11) 6 + 4
(12) 8 + 6
(13) 10 + 6
(14) 6 + 3
(15) 7 + 6
(16) 0 + 6
(17) 6 + 1
(18) 5 + 6
(19) 6 + 8
(20) 6 + 6
(21) 6 + 4
(22) 3 + 6
(23) 10 + 6
(24) 8 + 6
(25) 6 + 0
(26) 6 + 5
(27) 3 + 6
(28) 6 + 8
(29) 0 + 6
(30) 6 + 10
(31) 2 + 6
(32) 6 + 9
(33) 4 + 6
(34) 6 + 5
(35) 9 + 6
(36) 1 + 6
(37) 6 + 7
(38) 8 + 6
(39) 6 + 4
(40) 5 + 6
(41) 7 + 6
(42) 6 + 2
(43) 6 + 10
(44) 4 + 6
(45) 6 + 5
(46) 6 + 6
(47) 6 + 1
(48) 3 + 6
(49) 6 + 9
(50) 2 + 6
(51) 8 + 6
(52) 6 + 10
(53) 1 + 6
(54) 6 + 4
(55) 6 + 2
(56) 9 + 6
(57) 6 + 3
(58) 6 + 0
(59) 6 + 9
(60) 0 + 6
(61) 6 + 6
(62) 6 + 10
(63) 5 + 6
(64) 8 + 6
(65) 6 + 7
(66) 6 + 5
(67) 10 + 6
(68) 9 + 6
(69) 2 + 6
(70) 6 + 8
(71) 6 + 1
(72) 4 + 6
(73) 7 + 6
(74) 6 + 10
(75) 8 + 6
(76) 6 + 7
(77) 6 + 5
(78) 9 + 6
(79) 6 + 6
(80) 8 + 6
(81) 5 + 6
(82) 6 + 9
(83) 3 + 6
(84) 0 + 6
(85) 5 + 6
(86) 6 + 6
(87) 6 + 3
(88) 8 + 6
(89) 6 + 4
(90) 2 + 6
(91) 6 + 10
(92) 6 + 2
(93) 7 + 6
(94) 6 + 1
(95) 6 + 8
(96) 4 + 6
(97) 10 + 6
(98) 6 + 0
(99) 6 + 7
(100) 1 + 6

Addition 6

KEY

(1)	4 + 6	10	(26)	6 + 5	11	(51)	8 + 6	14	(76)	6 + 7	13
(2)	6 + 7	13	(27)	3 + 6	9	(52)	6 + 10	16	(77)	6 + 5	11
(3)	2 + 6	8	(28)	6 + 8	14	(53)	1 + 6	7	(78)	9 + 6	15
(4)	6 + 5	11	(29)	0 + 6	6	(54)	6 + 4	10	(79)	6 + 6	12
(5)	1 + 6	7	(30)	6 + 10	16	(55)	6 + 2	8	(80)	8 + 6	14
(6)	6 + 9	15	(31)	2 + 6	8	(56)	9 + 6	15	(81)	5 + 6	11
(7)	6 + 2	8	(32)	6 + 9	15	(57)	6 + 3	9	(82)	6 + 9	15
(8)	9 + 6	15	(33)	4 + 6	10	(58)	6 + 0	6	(83)	3 + 6	9
(9)	1 + 6	7	(34)	6 + 5	11	(59)	6 + 9	15	(84)	0 + 6	6
(10)	6 + 10	16	(35)	9 + 6	15	(60)	0 + 6	6	(85)	5 + 6	11
(11)	6 + 4	10	(36)	1 + 6	7	(61)	6 + 6	12	(86)	6 + 6	12
(12)	8 + 6	14	(37)	6 + 7	13	(62)	6 + 10	16	(87)	6 + 3	9
(13)	10 + 6	16	(38)	8 + 6	14	(63)	5 + 6	11	(88)	8 + 6	14
(14)	6 + 3	9	(39)	6 + 4	10	(64)	8 + 6	14	(89)	6 + 4	10
(15)	7 + 6	13	(40)	5 + 6	11	(65)	6 + 7	13	(90)	2 + 6	8
(16)	0 + 6	6	(41)	7 + 6	13	(66)	6 + 5	11	(91)	6 + 10	16
(17)	6 + 1	7	(42)	6 + 2	8	(67)	10 + 6	16	(92)	6 + 2	8
(18)	5 + 6	11	(43)	6 + 10	16	(68)	9 + 6	15	(93)	7 + 6	13
(19)	6 + 8	14	(44)	4 + 6	10	(69)	2 + 6	8	(94)	6 + 1	7
(20)	6 + 6	12	(45)	6 + 5	11	(70)	6 + 8	14	(95)	6 + 8	14
(21)	6 + 4	10	(46)	6 + 6	12	(71)	6 + 1	7	(96)	4 + 6	10
(22)	3 + 6	9	(47)	6 + 1	7	(72)	4 + 6	10	(97)	10 + 6	16
(23)	10 + 6	16	(48)	3 + 6	9	(73)	7 + 6	13	(98)	6 + 0	6
(24)	8 + 6	14	(49)	6 + 9	15	(74)	6 + 10	16	(99)	6 + 7	13
(25)	6 + 0	6	(50)	2 + 6	8	(75)	8 + 6	14	(100)	1 + 6	7

Addition 0-6

Name ______________________________ Date _______________ Score ______ Time ______

(1) $5+10$

(2) $3+9$

(3) $3+4$

(4) $5+2$

(5) $4+4$

(6) $3+7$

(7) $2+2$

(8) $3+0$

(9) $6+8$

(10) $5+6$

(11) $6+2$

(12) $1+3$

(13) $5+8$

(14) $6+9$

(15) $4+0$

(16) $5+4$

(17) $6+7$

(18) $4+7$

(19) $6+5$

(20) $1+5$

(21) $0+7$

(22) $5+3$

(23) $4+9$

(24) $3+8$

(25) $6+0$

(26) $2+2$

(27) $4+6$

(28) $0+5$

(29) $3+2$

(30) $6+1$

(31) $5+7$

(32) $6+6$

(33) $4+8$

(34) $6+10$

(35) $2+3$

(36) $6+4$

(37) $0+9$

(38) $6+1$

(39) $5+5$

(40) $6+3$

(41) $4+3$

(42) $2+8$

(43) $6+3$

(44) $5+9$

(45) $0+8$

(46) $2+4$

(47) $4+1$

(48) $6+4$

(49) $2+6$

(50) $6+8$

(51) $3+3$

(52) $4+6$

(53) $0+5$

(54) $6+8$

(55) $6+0$

(56) $2+6$

(57) $4+1$

(58) $6+4$

(59) $0+8$

(60) $2+4$

(61) $3+6$

(62) $5+9$

(63) $4+3$

(64) $2+8$

(65) $5+5$

(66) $6+3$

(67) $0+9$

(68) $6+1$

(69) $2+3$

(70) $6+4$

(71) $4+8$

(72) $6+10$

(73) $5+7$

(74) $6+6$

(75) $6+1$

(76) $8+6$

(77) $6+2$

(78) $5+6$

(79) $1+3$

(80) $6+9$

(81) $5+8$

(82) $5+4$

(83) $4+0$

(84) $4+7$

(85) $7+6$

(86) $1+5$

(87) $6+5$

(88) $5+3$

(89) $0+7$

(90) $3+8$

(91) $4+9$

(92) $5+10$

(93) $6+0$

(94) $3+4$

(95) $9+3$

(96) $4+4$

(97) $5+2$

(98) $2+2$

(99) $3+7$

(100) $3+0$

Addition 0-6

KEY

(1)	$5+10$	15	(26)	$2+2$	4	(51)	$3+3$	6	(76)	$8+6$	14
(2)	$3+9$	12	(27)	$4+6$	10	(52)	$4+6$	10	(77)	$6+2$	8
(3)	$3+4$	7	(28)	$0+5$	5	(53)	$0+5$	5	(78)	$5+6$	11
(4)	$5+2$	7	(29)	$3+2$	5	(54)	$6+8$	14	(79)	$1+3$	4
(5)	$4+4$	8	(30)	$6+1$	7	(55)	$6+0$	6	(80)	$6+9$	15
(6)	$3+7$	10	(31)	$5+7$	12	(56)	$2+6$	8	(81)	$5+8$	13
(7)	$2+2$	4	(32)	$6+6$	12	(57)	$4+1$	5	(82)	$5+4$	9
(8)	$3+0$	3	(33)	$4+8$	12	(58)	$6+4$	10	(83)	$4+0$	4
(9)	$6+8$	14	(34)	$6+10$	16	(59)	$0+8$	8	(84)	$4+7$	11
(10)	$5+6$	11	(35)	$2+3$	5	(60)	$2+4$	6	(85)	$7+6$	13
(11)	$6+2$	8	(36)	$6+4$	10	(61)	$3+6$	9	(86)	$1+5$	6
(12)	$1+3$	4	(37)	$0+9$	9	(62)	$5+9$	14	(87)	$6+5$	11
(13)	$5+8$	13	(38)	$6+1$	7	(63)	$4+3$	7	(88)	$5+3$	8
(14)	$6+9$	15	(39)	$5+5$	10	(64)	$2+8$	10	(89)	$0+7$	7
(15)	$4+0$	4	(40)	$6+3$	9	(65)	$5+5$	10	(90)	$3+8$	11
(16)	$5+4$	9	(41)	$4+3$	7	(66)	$6+3$	9	(91)	$4+9$	13
(17)	$6+7$	13	(42)	$2+8$	10	(67)	$0+9$	9	(92)	$5+10$	15
(18)	$4+7$	11	(43)	$6+3$	9	(68)	$6+1$	7	(93)	$6+0$	6
(19)	$6+5$	11	(44)	$5+9$	14	(69)	$2+3$	5	(94)	$3+4$	7
(20)	$1+5$	6	(45)	$0+8$	8	(70)	$6+4$	10	(95)	$9+3$	12
(21)	$0+7$	7	(46)	$2+4$	6	(71)	$4+8$	12	(96)	$4+4$	8
(22)	$5+3$	8	(47)	$4+1$	5	(72)	$6+10$	16	(97)	$5+2$	7
(23)	$4+9$	13	(48)	$6+4$	10	(73)	$5+7$	12	(98)	$2+2$	4
(24)	$3+8$	11	(49)	$2+6$	8	(74)	$6+6$	12	(99)	$3+7$	10
(25)	$6+0$	6	(50)	$6+8$	14	(75)	$6+1$	7	(100)	$3+0$	3

Addition 7

Name ______________________ Date ____________ Score _____ Time _____

(1) 6 + 7
(2) 7 + 0
(3) 4 + 7
(4) 10 + 7
(5) 7 + 6
(6) 7 + 8
(7) 9 + 7
(8) 1 + 7
(9) 7 + 7
(10) 7 + 3
(11) 7 + 1
(12) 0 + 7
(13) 6 + 7
(14) 7 + 5
(15) 7 + 2
(16) 7 + 9
(17) 5 + 7
(18) 7 + 10
(19) 3 + 7
(20) 8 + 7
(21) 7 + 5
(22) 0 + 7
(23) 7 + 2
(24) 7 + 4
(25) 7 + 6
(26) 3 + 7
(27) 7 + 9
(28) 7 + 1
(29) 7 + 6
(30) 1 + 7
(31) 7 + 0
(32) 7 + 4
(33) 8 + 7
(34) 5 + 7
(35) 7 + 4
(36) 7 + 9
(37) 3 + 7
(38) 8 + 7
(39) 7 + 7
(40) 7 + 8
(41) 4 + 7
(42) 2 + 7
(43) 7 + 5
(44) 7 + 10
(45) 7 + 4
(46) 7 + 3
(47) 9 + 7
(48) 6 + 7
(49) 0 + 7
(50) 10 + 7
(51) 8 + 7
(52) 2 + 7
(53) 10 + 7
(54) 7 + 7
(55) 4 + 7
(56) 7 + 0
(57) 1 + 7
(58) 7 + 9
(59) 0 + 7
(60) 10 + 7
(61) 7 + 3
(62) 7 + 10
(63) 2 + 7
(64) 7 + 8
(65) 5 + 7
(66) 7 + 4
(67) 6 + 7
(68) 7 + 1
(69) 3 + 7
(70) 6 + 7
(71) 7 + 9
(72) 2 + 7
(73) 7 + 5
(74) 4 + 7
(75) 7 + 7
(76) 5 + 7
(77) 7 + 1
(78) 7 + 3
(79) 9 + 7
(80) 1 + 7
(81) 10 + 7
(82) 7 + 2
(83) 7 + 4
(84) 3 + 7
(85) 2 + 7
(86) 7 + 7
(87) 4 + 7
(88) 9 + 7
(89) 7 + 1
(90) 5 + 7
(91) 0 + 7
(92) 7 + 10
(93) 7 + 5
(94) 7 + 0
(95) 7 + 8
(96) 6 + 7
(97) 7 + 6
(98) 7 + 9
(99) 7 + 3
(100) 8 + 7

Addition 7

KEY

No.	Problem	Answer
(1)	6 + 7	13
(2)	7 + 0	7
(3)	4 + 7	11
(4)	10 + 7	17
(5)	7 + 6	13
(6)	7 + 8	15
(7)	9 + 7	16
(8)	1 + 7	8
(9)	7 + 7	14
(10)	7 + 3	10
(11)	7 + 1	8
(12)	0 + 7	7
(13)	6 + 7	13
(14)	7 + 5	12
(15)	7 + 2	9
(16)	7 + 9	16
(17)	5 + 7	12
(18)	7 + 10	17
(19)	3 + 7	10
(20)	8 + 7	15
(21)	7 + 5	12
(22)	0 + 7	7
(23)	7 + 2	9
(24)	7 + 4	11
(25)	7 + 6	13
(26)	3 + 7	10
(27)	7 + 9	16
(28)	7 + 1	8
(29)	7 + 6	13
(30)	1 + 7	8
(31)	7 + 0	7
(32)	7 + 4	11
(33)	8 + 7	15
(34)	5 + 7	12
(35)	7 + 4	11
(36)	7 + 9	16
(37)	3 + 7	10
(38)	8 + 7	15
(39)	7 + 7	14
(40)	7 + 8	15
(41)	4 + 7	11
(42)	2 + 7	9
(43)	7 + 5	12
(44)	7 + 10	17
(45)	7 + 4	11
(46)	7 + 3	10
(47)	9 + 7	16
(48)	6 + 7	13
(49)	0 + 7	7
(50)	10 + 7	17
(51)	8 + 7	15
(52)	2 + 7	9
(53)	10 + 7	17
(54)	7 + 7	14
(55)	4 + 7	11
(56)	7 + 0	7
(57)	1 + 7	8
(58)	7 + 9	16
(59)	0 + 7	7
(60)	10 + 7	17
(61)	7 + 3	10
(62)	7 + 10	17
(63)	2 + 7	9
(64)	7 + 8	15
(65)	5 + 7	12
(66)	7 + 4	11
(67)	6 + 7	13
(68)	7 + 1	8
(69)	3 + 7	10
(70)	6 + 7	13
(71)	7 + 9	16
(72)	2 + 7	9
(73)	7 + 5	12
(74)	4 + 7	11
(75)	7 + 7	14
(76)	5 + 7	12
(77)	7 + 1	8
(78)	7 + 3	10
(79)	9 + 7	16
(80)	1 + 7	8
(81)	10 + 7	17
(82)	7 + 2	9
(83)	7 + 4	11
(84)	3 + 7	10
(85)	2 + 7	9
(86)	7 + 7	14
(87)	4 + 7	11
(88)	9 + 7	16
(89)	7 + 1	8
(90)	5 + 7	12
(91)	0 + 7	7
(92)	7 + 10	17
(93)	7 + 5	12
(94)	7 + 0	7
(95)	7 + 8	15
(96)	6 + 7	13
(97)	7 + 6	13
(98)	7 + 9	16
(99)	7 + 3	10
(100)	8 + 7	15

Addition 0-7

Name ______________________________ Date _______________ Score ______ Time ______

(1) 5 + 6
(2) 5 + 3
(3) 4 + 4
(4) 7 + 5
(5) 6 + 10
(6) 2 + 8
(7) 7 + 4
(8) 4 + 1
(9) 3 + 6
(10) 7 + 9
(11) 6 + 1
(12) 5 + 2
(13) 2 + 3
(14) 3 + 1
(15) 7 + 0
(16) 7 + 3
(17) 4 + 10
(18) 5 + 8
(19) 4 + 3
(20) 2 + 4
(21) 5 + 10
(22) 7 + 1
(23) 6 + 5
(24) 4 + 8
(25) 2 + 7
(26) 7 + 8
(27) 6 + 7
(28) 7 + 10
(29) 4 + 5
(30) 3 + 3
(31) 2 + 7
(32) 2 + 4
(33) 3 + 5
(34) 7 + 7
(35) 5 + 7
(36) 6 + 9
(37) 2 + 10
(38) 6 + 4
(39) 7 + 1
(40) 7 + 9
(41) 5 + 6
(42) 2 + 9
(43) 3 + 7
(44) 4 + 8
(45) 7 + 0
(46) 6 + 8
(47) 7 + 9
(48) 5 + 5
(49) 4 + 9
(50) 7 + 10
(51) 3 + 5
(52) 4 + 7
(53) 6 + 2
(54) 2 + 5
(55) 7 + 7
(56) 4 + 9
(57) 2 + 10
(58) 5 + 5
(59) 4 + 2
(60) 7 + 10
(61) 7 + 2
(62) 6 + 8
(63) 3 + 3
(64) 5 + 9
(65) 3 + 10
(66) 7 + 6
(67) 6 + 4
(68) 5 + 7
(69) 3 + 8
(70) 2 + 6
(71) 2 + 2
(72) 6 + 6
(73) 7 + 8
(74) 3 + 7
(75) 4 + 5
(76) 7 + 10
(77) 5 + 5
(78) 4 + 9
(79) 7 + 9
(80) 6 + 8
(81) 3 + 7
(82) 7 + 0
(83) 4 + 8
(84) 2 + 9
(85) 7 + 1
(86) 5 + 6
(87) 7 + 9
(88) 6 + 4
(89) 5 + 7
(90) 6 + 9
(91) 2 + 10
(92) 7 + 7
(93) 2 + 4
(94) 3 + 5
(95) 2 + 7
(96) 3 + 3
(97) 7 + 10
(98) 4 + 5
(99) 6 + 7
(100) 7 + 8

Addition 0-7

KEY

No.	Problem	Answer
(1)	5 + 6	11
(2)	5 + 3	8
(3)	4 + 4	8
(4)	7 + 5	12
(5)	6 + 10	16
(6)	2 + 8	10
(7)	7 + 4	11
(8)	4 + 1	5
(9)	3 + 6	9
(10)	7 + 9	16
(11)	6 + 1	7
(12)	5 + 2	7
(13)	2 + 3	5
(14)	3 + 1	4
(15)	7 + 0	7
(16)	7 + 3	10
(17)	4 + 10	14
(18)	5 + 8	13
(19)	4 + 3	7
(20)	2 + 4	6
(21)	5 + 10	15
(22)	7 + 1	8
(23)	6 + 5	11
(24)	4 + 8	12
(25)	2 + 7	9
(26)	7 + 8	15
(27)	6 + 7	13
(28)	7 + 10	17
(29)	4 + 5	9
(30)	3 + 3	6
(31)	2 + 7	9
(32)	2 + 4	6
(33)	3 + 5	8
(34)	7 + 7	14
(35)	5 + 7	12
(36)	6 + 9	15
(37)	2 + 10	12
(38)	6 + 4	10
(39)	7 + 1	8
(40)	7 + 9	16
(41)	5 + 6	11
(42)	2 + 9	11
(43)	3 + 7	10
(44)	4 + 8	12
(45)	7 + 0	7
(46)	6 + 8	14
(47)	7 + 9	16
(48)	5 + 5	10
(49)	4 + 9	13
(50)	7 + 10	17
(51)	3 + 5	8
(52)	4 + 7	11
(53)	6 + 2	8
(54)	2 + 5	7
(55)	7 + 7	14
(56)	4 + 9	13
(57)	2 + 10	12
(58)	5 + 5	10
(59)	4 + 2	6
(60)	7 + 10	17
(61)	7 + 2	9
(62)	6 + 8	14
(63)	3 + 3	6
(64)	5 + 9	14
(65)	3 + 10	13
(66)	7 + 6	13
(67)	6 + 4	10
(68)	5 + 7	12
(69)	3 + 8	11
(70)	2 + 6	8
(71)	2 + 2	4
(72)	6 + 6	12
(73)	7 + 8	15
(74)	3 + 7	10
(75)	4 + 5	9
(76)	7 + 10	17
(77)	5 + 5	10
(78)	4 + 9	13
(79)	7 + 9	16
(80)	6 + 8	14
(81)	3 + 7	10
(82)	7 + 0	7
(83)	4 + 8	12
(84)	2 + 9	11
(85)	7 + 1	8
(86)	5 + 6	11
(87)	7 + 9	16
(88)	6 + 4	10
(89)	5 + 7	12
(90)	6 + 9	15
(91)	2 + 10	12
(92)	7 + 7	14
(93)	2 + 4	6
(94)	3 + 5	8
(95)	2 + 7	9
(96)	3 + 3	6
(97)	7 + 10	17
(98)	4 + 5	9
(99)	6 + 7	13
(100)	7 + 8	15

Addition 8

Name ______________________________ Date _______________ Score ______ Time ______

(1) 9 + 8
(2) 0 + 8
(3) 3 + 8
(4) 10 + 8
(5) 8 + 0
(6) 4 + 8
(7) 6 + 8
(8) 1 + 8
(9) 8 + 10
(10) 8 + 8
(11) 5 + 8
(12) 8 + 9
(13) 8 + 1
(14) 8 + 7
(15) 2 + 8
(16) 7 + 8
(17) 8 + 6
(18) 8 + 2
(19) 8 + 4
(20) 8 + 5
(21) 2 + 8
(22) 7 + 8
(23) 3 + 8
(24) 0 + 8
(25) 8 + 3
(26) 8 + 4
(27) 9 + 8
(28) 8 + 3
(29) 8 + 7
(30) 8 + 2
(31) 8 + 9
(32) 8 + 1
(33) 7 + 8
(34) 0 + 8
(35) 8 + 7
(36) 2 + 8
(37) 8 + 0
(38) 5 + 8
(39) 8 + 6
(40) 3 + 8
(41) 8 + 8
(42) 8 + 2
(43) 10 + 8
(44) 8 + 9
(45) 8 + 4
(46) 8 + 10
(47) 4 + 8
(48) 8 + 5
(49) 6 + 8
(50) 1 + 8
(51) 8 + 8
(52) 8 + 10
(53) 8 + 0
(54) 3 + 8
(55) 0 + 8
(56) 7 + 8
(57) 5 + 8
(58) 2 + 8
(59) 8 + 1
(60) 8 + 3
(61) 9 + 8
(62) 8 + 5
(63) 8 + 10
(64) 6 + 8
(65) 4 + 8
(66) 1 + 8
(67) 8 + 4
(68) 8 + 5
(69) 8 + 9
(70) 10 + 8
(71) 8 + 8
(72) 8 + 6
(73) 8 + 2
(74) 8 + 7
(75) 4 + 8
(76) 8 + 5
(77) 7 + 8
(78) 8 + 10
(79) 0 + 8
(80) 4 + 8
(81) 8 + 7
(82) 8 + 4
(83) 8 + 2
(84) 8 + 9
(85) 8 + 0
(86) 10 + 8
(87) 5 + 8
(88) 2 + 8
(89) 8 + 6
(90) 8 + 8
(91) 3 + 8
(92) 8 + 9
(93) 10 + 8
(94) 5 + 8
(95) 8 + 3
(96) 6 + 8
(97) 1 + 8
(98) 9 + 8
(99) 8 + 6
(100) 8 + 1

Addition 8

KEY

(1)	9 + 8	17	(26)	8 + 4	12	(51)	8 + 8	16	(76)	8 + 5	13
(2)	0 + 8	8	(27)	9 + 8	17	(52)	8 + 10	18	(77)	7 + 8	15
(3)	3 + 8	11	(28)	8 + 3	11	(53)	8 + 0	8	(78)	8 + 10	18
(4)	10 + 8	18	(29)	8 + 7	15	(54)	3 + 8	11	(79)	0 + 8	8
(5)	8 + 0	8	(30)	8 + 2	10	(55)	0 + 8	8	(80)	4 + 8	12
(6)	4 + 8	12	(31)	8 + 9	17	(56)	7 + 8	15	(81)	8 + 7	15
(7)	6 + 8	14	(32)	8 + 1	9	(57)	5 + 8	13	(82)	8 + 4	12
(8)	1 + 8	9	(33)	7 + 8	15	(58)	2 + 8	10	(83)	8 + 2	10
(9)	8 + 10	18	(34)	0 + 8	8	(59)	8 + 1	9	(84)	8 + 9	17
(10)	8 + 8	16	(35)	8 + 7	15	(60)	8 + 3	11	(85)	8 + 0	8
(11)	5 + 8	13	(36)	2 + 8	10	(61)	9 + 8	17	(86)	10 + 8	18
(12)	8 + 9	17	(37)	8 + 0	8	(62)	8 + 5	13	(87)	5 + 8	13
(13)	8 + 1	9	(38)	5 + 8	13	(63)	8 + 10	18	(88)	2 + 8	10
(14)	8 + 7	15	(39)	8 + 6	14	(64)	6 + 8	14	(89)	8 + 6	14
(15)	2 + 8	10	(40)	3 + 8	11	(65)	4 + 8	12	(90)	8 + 8	16
(16)	7 + 8	15	(41)	8 + 8	16	(66)	1 + 8	9	(91)	3 + 8	11
(17)	8 + 6	14	(42)	8 + 2	10	(67)	8 + 4	12	(92)	8 + 9	17
(18)	8 + 2	10	(43)	10 + 8	18	(68)	8 + 5	13	(93)	10 + 8	18
(19)	8 + 4	12	(44)	8 + 9	17	(69)	8 + 9	17	(94)	5 + 8	13
(20)	8 + 5	13	(45)	8 + 4	12	(70)	10 + 8	18	(95)	8 + 3	11
(21)	2 + 8	10	(46)	8 + 10	18	(71)	8 + 8	16	(96)	6 + 8	14
(22)	7 + 8	15	(47)	4 + 8	12	(72)	8 + 6	14	(97)	1 + 8	9
(23)	3 + 8	11	(48)	8 + 5	13	(73)	8 + 2	10	(98)	9 + 8	17
(24)	0 + 8	8	(49)	6 + 8	14	(74)	8 + 7	15	(99)	8 + 6	14
(25)	8 + 3	11	(50)	1 + 8	9	(75)	4 + 8	12	(100)	8 + 1	9

Addition 0-8

Name____________________________ Date _______________ Score ______ Time______

(1) 4 + 3
(2) 8 + 7
(3) 3 + 8
(4) 6 + 1
(5) 4 + 2
(6) 8 + 3
(7) 2 + 5
(8) 8 + 5
(9) 3 + 2
(10) 5 + 0
(11) 6 + 2
(12) 4 + 4
(13) 1 + 5
(14) 6 + 9
(15) 4 + 10
(16) 8 + 6
(17) 4 + 4
(18) 4 + 1
(19) 8 + 9
(20) 6 + 5
(21) 3 + 5
(22) 3 + 8
(23) 7 + 3
(24) 5 + 2
(25) 8 + 4
(26) 2 + 5
(27) 6 + 0
(28) 6 + 5
(29) 0 + 10
(30) 7 + 7
(31) 0 + 5
(32) 8 + 2
(33) 3 + 0
(34) 1 + 10
(35) 9 + 1
(36) 0 + 0
(37) 2 + 6
(38) 8 + 3
(39) 6 + 6
(40) 2 + 8
(41) 3 + 1
(42) 8 + 7
(43) 2 + 2
(44) 1 + 1
(45) 4 + 8
(46) 3 + 4
(47) 8 + 1
(48) 2 + 0
(49) 8 + 8
(50) 6 + 9
(51) 2 + 3
(52) 8 + 2
(53) 2 + 1
(54) 3 + 10
(55) 7 + 9
(56) 8 + 0
(57) 6 + 3
(58) 4 + 9
(59) 5 + 10
(60) 3 + 2
(61) 5 + 6
(62) 4 + 0
(63) 5 + 5
(64) 2 + 2
(65) 2 + 10
(66) 6 + 6
(67) 5 + 3
(68) 2 + 9
(69) 6 + 10
(70) 7 + 1
(71) 3 + 6
(72) 5 + 4
(73) 6 + 2
(74) 2 + 8
(75) 1 + 8
(76) 5 + 7
(77) 8 + 9
(78) 5 + 9
(79) 7 + 8
(80) 5 + 10
(81) 8 + 4
(82) 7 + 5
(83) 5 + 2
(84) 8 + 1
(85) 8 + 6
(86) 1 + 3
(87) 8 + 5
(88) 6 + 7
(89) 8 + 10
(90) 7 + 2
(91) 6 + 4
(92) 6 + 8
(93) 7 + 3
(94) 8 + 0
(95) 4 + 6
(96) 5 + 8
(97) 4 + 2
(98) 3 + 9
(99) 8 + 10
(100) 8 + 8

Addition 0-8

KEY

(1)	4 + 3	7	(26)	2 + 5	7	(51)	2 + 3	5	(76)	5 + 7	12
(2)	8 + 7	15	(27)	6 + 0	6	(52)	8 + 2	10	(77)	8 + 9	17
(3)	3 + 8	11	(28)	6 + 5	11	(53)	2 + 1	3	(78)	5 + 9	14
(4)	6 + 1	7	(29)	0 + 10	10	(54)	3 + 10	13	(79)	7 + 8	15
(5)	4 + 2	6	(30)	7 + 7	14	(55)	7 + 9	16	(80)	5 + 10	15
(6)	8 + 3	11	(31)	0 + 5	5	(56)	8 + 0	8	(81)	8 + 4	12
(7)	2 + 5	7	(32)	8 + 2	10	(57)	6 + 3	9	(82)	7 + 5	12
(8)	8 + 5	13	(33)	3 + 0	3	(58)	4 + 9	13	(83)	5 + 2	7
(9)	3 + 2	5	(34)	1 + 10	11	(59)	5 + 10	15	(84)	8 + 1	9
(10)	5 + 0	5	(35)	9 + 1	10	(60)	3 + 2	5	(85)	8 + 6	14
(11)	6 + 2	8	(36)	0 + 0	0	(61)	5 + 6	11	(86)	1 + 3	4
(12)	4 + 4	8	(37)	2 + 6	8	(62)	4 + 0	4	(87)	8 + 5	13
(13)	1 + 5	6	(38)	8 + 3	11	(63)	5 + 5	10	(88)	6 + 7	13
(14)	6 + 9	15	(39)	6 + 6	12	(64)	2 + 2	4	(89)	8 + 10	18
(15)	4 + 10	14	(40)	2 + 8	10	(65)	2 + 10	12	(90)	7 + 2	9
(16)	8 + 6	14	(41)	3 + 1	4	(66)	6 + 6	12	(91)	6 + 4	10
(17)	4 + 4	8	(42)	8 + 7	15	(67)	5 + 3	8	(92)	6 + 8	14
(18)	4 + 1	5	(43)	2 + 2	4	(68)	2 + 9	11	(93)	7 + 3	10
(19)	8 + 9	17	(44)	1 + 1	2	(69)	6 + 10	16	(94)	8 + 0	8
(20)	6 + 5	11	(45)	4 + 8	12	(70)	7 + 1	8	(95)	4 + 6	10
(21)	3 + 5	8	(46)	3 + 4	7	(71)	3 + 6	9	(96)	5 + 8	13
(22)	3 + 8	11	(47)	8 + 1	9	(72)	5 + 4	9	(97)	4 + 2	6
(23)	7 + 3	10	(48)	2 + 0	2	(73)	6 + 2	8	(98)	3 + 9	12
(24)	5 + 2	7	(49)	8 + 8	16	(74)	2 + 8	10	(99)	8 + 10	18
(25)	8 + 4	12	(50)	6 + 9	15	(75)	1 + 8	9	(100)	8 + 8	16

Addition 9

Name ______________________ Date ______________ Score ______ Time ______

(1) 9 + 8
(2) 9 + 6
(3) 3 + 9
(4) 7 + 9
(5) 10 + 9
(6) 4 + 9
(7) 1 + 9
(8) 9 + 8
(9) 9 + 0
(10) 9 + 2
(11) 7 + 9
(12) 9 + 6
(13) 0 + 9
(14) 3 + 9
(15) 9 + 5
(16) 9 + 9
(17) 9 + 3
(18) 6 + 9
(19) 9 + 1
(20) 8 + 9
(21) 5 + 9
(22) 9 + 10
(23) 2 + 9
(24) 9 + 4
(25) 9 + 7

(26) 2 + 9
(27) 9 + 10
(28) 6 + 9
(29) 1 + 9
(30) 9 + 9
(31) 0 + 9
(32) 9 + 2
(33) 1 + 9
(34) 5 + 9
(35) 9 + 4
(36) 9 + 7
(37) 5 + 9
(38) 6 + 9
(39) 9 + 1
(40) 9 + 5
(41) 9 + 6
(42) 9 + 0
(43) 4 + 9
(44) 10 + 9
(45) 9 + 8
(46) 7 + 9
(47) 3 + 9
(48) 9 + 3
(49) 8 + 9
(50) 2 + 9

(51) 8 + 9
(52) 9 + 7
(53) 10 + 9
(54) 4 + 9
(55) 0 + 9
(56) 1 + 9
(57) 9 + 4
(58) 5 + 9
(59) 9 + 1
(60) 9 + 6
(61) 4 + 9
(62) 9 + 8
(63) 3 + 9
(64) 8 + 9
(65) 2 + 9
(66) 9 + 3
(67) 7 + 9
(68) 10 + 9
(69) 9 + 0
(70) 9 + 5
(71) 6 + 9
(72) 9 + 7
(73) 5 + 9
(74) 9 + 2
(75) 9 + 9

(76) 5 + 9
(77) 0 + 9
(78) 9 + 9
(79) 9 + 4
(80) 9 + 7
(81) 2 + 9
(82) 5 + 9
(83) 9 + 1
(84) 9 + 3
(85) 9 + 5
(86) 0 + 9
(87) 7 + 9
(88) 9 + 0
(89) 1 + 9
(90) 10 + 9
(91) 4 + 9
(92) 9 + 8
(93) 9 + 2
(94) 9 + 6
(95) 3 + 9
(96) 9 + 9
(97) 6 + 9
(98) 8 + 9
(99) 9 + 10
(100) 9 + 4

Addition 9

KEY

(1)	9 + 8	17	(26)	2 + 9	11	(51)	8 + 9	17	(76)	5 + 9	14
(2)	9 + 6	15	(27)	9 + 10	19	(52)	9 + 7	16	(77)	0 + 9	9
(3)	3 + 9	12	(28)	6 + 9	15	(53)	10 + 9	19	(78)	9 + 9	18
(4)	7 + 9	16	(29)	1 + 9	10	(54)	4 + 9	13	(79)	9 + 4	13
(5)	10 + 9	19	(30)	9 + 9	18	(55)	0 + 9	9	(80)	9 + 7	16
(6)	4 + 9	13	(31)	0 + 9	9	(56)	1 + 9	10	(81)	2 + 9	11
(7)	1 + 9	10	(32)	9 + 2	11	(57)	9 + 4	13	(82)	5 + 9	14
(8)	9 + 8	17	(33)	1 + 9	10	(58)	5 + 9	14	(83)	9 + 1	10
(9)	9 + 0	9	(34)	5 + 9	14	(59)	9 + 1	10	(84)	9 + 3	12
(10)	9 + 2	11	(35)	9 + 4	13	(60)	9 + 6	15	(85)	9 + 5	14
(11)	7 + 9	16	(36)	9 + 7	16	(61)	4 + 9	13	(86)	0 + 9	9
(12)	9 + 6	15	(37)	5 + 9	14	(62)	9 + 8	17	(87)	7 + 9	16
(13)	0 + 9	9	(38)	6 + 9	15	(63)	3 + 9	12	(88)	9 + 0	9
(14)	3 + 9	12	(39)	9 + 1	10	(64)	8 + 9	17	(89)	1 + 9	10
(15)	9 + 5	14	(40)	9 + 5	14	(65)	2 + 9	11	(90)	10 + 9	19
(16)	9 + 9	18	(41)	9 + 6	15	(66)	9 + 3	12	(91)	4 + 9	13
(17)	9 + 3	12	(42)	9 + 0	9	(67)	7 + 9	16	(92)	9 + 8	17
(18)	6 + 9	15	(43)	4 + 9	13	(68)	10 + 9	19	(93)	9 + 2	11
(19)	9 + 1	10	(44)	10 + 9	19	(69)	9 + 0	9	(94)	9 + 6	15
(20)	8 + 9	17	(45)	9 + 8	17	(70)	9 + 5	14	(95)	3 + 9	12
(21)	5 + 9	14	(46)	7 + 9	16	(71)	6 + 9	15	(96)	9 + 9	18
(22)	9 + 10	19	(47)	3 + 9	12	(72)	9 + 7	16	(97)	6 + 9	15
(23)	2 + 9	11	(48)	9 + 3	12	(73)	5 + 9	14	(98)	8 + 9	17
(24)	9 + 4	13	(49)	8 + 9	17	(74)	9 + 2	11	(99)	9 + 10	19
(25)	9 + 7	16	(50)	2 + 9	11	(75)	9 + 9	18	(100)	9 + 4	13

Addition 0-9

Name______________________ Date ______________ Score ______ Time______

(1) 9 + 6
(2) 2 + 7
(3) 8 + 4
(4) 7 + 9
(5) 3 + 2
(6) 9 + 5
(7) 7 + 8
(8) 4 + 3
(9) 9 + 0
(10) 7 + 3
(11) 3 + 10
(12) 9 + 8
(13) 4 + 6
(14) 2 + 8
(15) 6 + 7
(16) 7 + 10
(17) 9 + 4
(18) 7 + 5
(19) 5 + 3
(20) 9 + 1
(21) 2 + 10
(22) 8 + 7
(23) 3 + 3
(24) 0 + 7
(25) 5 + 5
(26) 2 + 3
(27) 3 + 9
(28) 7 + 4
(29) 1 + 5
(30) 7 + 7
(31) 9 + 3
(32) 5 + 8
(33) 8 + 2
(34) 6 + 9
(35) 8 + 0
(36) 9 + 7
(37) 3 + 6
(38) 9 + 2
(39) 8 + 8
(40) 9 + 0
(41) 8 + 5
(42) 5 + 5
(43) 9 + 10
(44) 5 + 6
(45) 8 + 10
(46) 0 + 6
(47) 9 + 9
(48) 4 + 2
(49) 9 + 4
(50) 3 + 9
(51) 9 + 10
(52) 6 + 4
(53) 4 + 10
(54) 9 + 6
(55) 3 + 4
(56) 6 + 6
(57) 2 + 5
(58) 5 + 9
(59) 5 + 2
(60) 3 + 9
(61) 7 + 6
(62) 9 + 8
(63) 4 + 5
(64) 9 + 2
(65) 7 + 10
(66) 4 + 7
(67) 8 + 1
(68) 5 + 4
(69) 5 + 7
(70) 9 + 4
(71) 0 + 4
(72) 2 + 4
(73) 1 + 2
(74) 9 + 7
(75) 0 + 10
(76) 9 + 0
(77) 1 + 6
(78) 8 + 3
(79) 9 + 1
(80) 5 + 10
(81) 3 + 8
(82) 2 + 9
(83) 8 + 6
(84) 6 + 3
(85) 8 + 9
(86) 4 + 4
(87) 6 + 8
(88) 9 + 9
(89) 3 + 5
(90) 6 + 2
(91) 6 + 5
(92) 1 + 9
(93) 4 + 8
(94) 9 + 2
(95) 5 + 1
(96) 3 + 7
(97) 2 + 6
(98) 2 + 2
(99) 9 + 5
(100) 7 + 2

Addition 0-9

KEY

(1)	9 + 6	15	(26)	2 + 3	5	(51)	9 + 10	19	(76)	9 + 0	9
(2)	2 + 7	9	(27)	3 + 9	12	(52)	6 + 4	10	(77)	1 + 6	7
(3)	8 + 4	12	(28)	7 + 4	11	(53)	4 + 10	14	(78)	8 + 3	11
(4)	7 + 9	16	(29)	1 + 5	6	(54)	9 + 6	15	(79)	9 + 1	10
(5)	3 + 2	5	(30)	7 + 7	14	(55)	3 + 4	7	(80)	5 + 10	15
(6)	9 + 5	14	(31)	9 + 3	12	(56)	6 + 6	12	(81)	3 + 8	11
(7)	7 + 8	15	(32)	5 + 8	13	(57)	2 + 5	7	(82)	2 + 9	11
(8)	4 + 3	7	(33)	8 + 2	10	(58)	5 + 9	14	(83)	8 + 6	14
(9)	9 + 0	9	(34)	6 + 9	15	(59)	5 + 2	7	(84)	6 + 3	9
(10)	7 + 3	10	(35)	8 + 0	8	(60)	3 + 9	12	(85)	8 + 9	17
(11)	3 + 10	13	(36)	9 + 7	16	(61)	7 + 6	13	(86)	4 + 4	8
(12)	9 + 8	17	(37)	3 + 6	9	(62)	9 + 8	17	(87)	6 + 8	14
(13)	4 + 6	10	(38)	9 + 2	11	(63)	4 + 5	9	(88)	9 + 9	18
(14)	2 + 8	10	(39)	8 + 8	16	(64)	9 + 2	11	(89)	3 + 5	8
(15)	6 + 7	13	(40)	9 + 0	9	(65)	7 + 10	17	(90)	6 + 2	8
(16)	7 + 10	17	(41)	8 + 5	13	(66)	4 + 7	11	(91)	6 + 5	11
(17)	9 + 4	13	(42)	5 + 5	10	(67)	8 + 1	9	(92)	1 + 9	10
(18)	7 + 5	12	(43)	9 + 10	19	(68)	5 + 4	9	(93)	4 + 8	12
(19)	5 + 3	8	(44)	5 + 6	11	(69)	5 + 7	12	(94)	9 + 2	11
(20)	9 + 1	10	(45)	8 + 10	18	(70)	9 + 4	13	(95)	5 + 1	6
(21)	2 + 10	12	(46)	0 + 6	6	(71)	0 + 4	4	(96)	3 + 7	10
(22)	8 + 7	15	(47)	9 + 9	18	(72)	2 + 4	6	(97)	2 + 6	8
(23)	3 + 3	6	(48)	4 + 2	6	(73)	1 + 2	3	(98)	2 + 2	4
(24)	0 + 7	7	(49)	9 + 4	13	(74)	9 + 7	16	(99)	9 + 5	14
(25)	5 + 5	10	(50)	3 + 9	12	(75)	0 + 10	10	(100)	7 + 2	9

Addition 10

Name ______________________________ Date ______________ Score ______ Time ______

(1) $10 + 8$
(2) $10 + 1$
(3) $9 + 10$
(4) $10 + 5$
(5) $10 + 7$
(6) $2 + 10$
(7) $1 + 10$
(8) $10 + 4$
(9) $7 + 10$
(10) $4 + 10$
(11) $10 + 3$
(12) $10 + 6$
(13) $10 + 0$
(14) $3 + 10$
(15) $10 + 5$
(16) $8 + 10$
(17) $4 + 10$
(18) $5 + 10$
(19) $10 + 3$
(20) $0 + 10$
(21) $10 + 6$
(22) $10 + 9$
(23) $10 + 8$
(24) $3 + 10$
(25) $7 + 10$
(26) $0 + 10$
(27) $10 + 9$
(28) $6 + 10$
(29) $10 + 2$
(30) $5 + 10$
(31) $10 + 10$
(32) $8 + 10$
(33) $1 + 10$
(34) $10 + 7$
(35) $2 + 10$
(36) $10 + 5$
(37) $9 + 10$
(38) $10 + 1$
(39) $10 + 8$
(40) $1 + 10$
(41) $2 + 10$
(42) $10 + 7$
(43) $10 + 5$
(44) $9 + 10$
(45) $10 + 1$
(46) $10 + 8$
(47) $10 + 5$
(48) $5 + 10$
(49) $2 + 10$
(50) $10 + 10$
(51) $10 + 6$
(52) $10 + 0$
(53) $3 + 10$
(54) $4 + 10$
(55) $7 + 10$
(56) $10 + 3$
(57) $0 + 10$
(58) $10 + 4$
(59) $8 + 10$
(60) $10 + 9$
(61) $5 + 10$
(62) $10 + 10$
(63) $10 + 2$
(64) $6 + 10$
(65) $10 + 3$
(66) $10 + 10$
(67) $7 + 10$
(68) $10 + 2$
(69) $6 + 10$
(70) $10 + 9$
(71) $0 + 10$
(72) $4 + 10$
(73) $8 + 10$
(74) $10 + 6$
(75) $10 + 3$
(76) $1 + 10$
(77) $10 + 4$
(78) $8 + 10$
(79) $10 + 10$
(80) $10 + 3$
(81) $2 + 10$
(82) $10 + 7$
(83) $10 + 2$
(84) $5 + 10$
(85) $7 + 10$
(86) $4 + 10$
(87) $10 + 5$
(88) $9 + 10$
(89) $3 + 10$
(90) $6 + 10$
(91) $10 + 9$
(92) $10 + 0$
(93) $10 + 1$
(94) $10 + 8$
(95) $10 + 6$
(96) $10 + 10$
(97) $9 + 10$
(98) $10 + 4$
(99) $10 + 7$
(100) $6 + 10$

Addition 10

KEY

(1)	$10+8$	18	(26)	$0+10$	10	(51)	$10+6$	16	(76)	$1+10$	11
(2)	$10+1$	11	(27)	$10+9$	19	(52)	$10+0$	10	(77)	$10+4$	14
(3)	$9+10$	19	(28)	$6+10$	16	(53)	$3+10$	13	(78)	$8+10$	18
(4)	$10+5$	15	(29)	$10+2$	12	(54)	$4+10$	14	(79)	$10+10$	20
(5)	$10+7$	17	(30)	$5+10$	15	(55)	$7+10$	17	(80)	$10+3$	13
(6)	$2+10$	12	(31)	$10+10$	20	(56)	$10+3$	13	(81)	$2+10$	12
(7)	$1+10$	11	(32)	$8+10$	18	(57)	$0+10$	10	(82)	$10+7$	17
(8)	$10+4$	14	(33)	$1+10$	11	(58)	$10+4$	14	(83)	$10+2$	12
(9)	$7+10$	17	(34)	$10+7$	17	(59)	$8+10$	18	(84)	$5+10$	15
(10)	$4+10$	14	(35)	$2+10$	12	(60)	$10+9$	19	(85)	$7+10$	17
(11)	$10+3$	13	(36)	$10+5$	15	(61)	$5+10$	15	(86)	$4+10$	14
(12)	$10+6$	16	(37)	$9+10$	19	(62)	$10+10$	20	(87)	$10+5$	15
(13)	$10+0$	10	(38)	$10+1$	11	(63)	$10+2$	12	(88)	$9+10$	19
(14)	$3+10$	13	(39)	$10+8$	18	(64)	$6+10$	16	(89)	$3+10$	13
(15)	$10+5$	15	(40)	$1+10$	11	(65)	$10+3$	13	(90)	$6+10$	16
(16)	$8+10$	18	(41)	$2+10$	12	(66)	$10+10$	20	(91)	$10+9$	19
(17)	$4+10$	14	(42)	$10+7$	17	(67)	$7+10$	17	(92)	$10+0$	10
(18)	$5+10$	15	(43)	$10+5$	15	(68)	$10+2$	12	(93)	$10+1$	11
(19)	$10+3$	13	(44)	$9+10$	19	(69)	$6+10$	16	(94)	$10+8$	18
(20)	$0+10$	10	(45)	$10+1$	11	(70)	$10+9$	19	(95)	$10+6$	16
(21)	$10+6$	16	(46)	$10+8$	18	(71)	$0+10$	10	(96)	$10+10$	20
(22)	$10+9$	19	(47)	$10+5$	15	(72)	$4+10$	14	(97)	$9+10$	19
(23)	$10+8$	18	(48)	$5+10$	15	(73)	$8+10$	18	(98)	$10+4$	14
(24)	$3+10$	13	(49)	$2+10$	12	(74)	$10+6$	16	(99)	$10+7$	17
(25)	$7+10$	17	(50)	$10+10$	20	(75)	$10+3$	13	(100)	$6+10$	16

Addition I

Name ______________________________ Date _______________ Score ______ Time ______

(1) $9+10$
(2) $8+3$
(3) $3+2$
(4) $6+2$
(5) $1+5$
(6) $0+2$
(7) $4+8$
(8) $3+9$
(9) $2+6$
(10) $0+10$
(11) $8+8$
(12) $9+7$
(13) $10+5$
(14) $6+3$
(15) $9+2$
(16) $8+1$
(17) $1+2$
(18) $0+1$
(19) $3+1$
(20) $6+4$
(21) $4+10$
(22) $9+4$
(23) $5+8$
(24) $5+7$
(25) $4+2$
(26) $6+10$
(27) $2+3$
(28) $2+5$
(29) $2+1$
(30) $3+8$
(31) $4+6$
(32) $7+7$
(33) $6+6$
(34) $8+5$
(35) $3+7$
(36) $1+3$
(37) $0+4$
(38) $2+4$
(39) $7+4$
(40) $6+8$
(41) $1+8$
(42) $10+3$
(43) $9+8$
(44) $5+10$
(45) $10+6$
(46) $7+9$
(47) $6+5$
(48) $4+5$
(49) $4+9$
(50) $9+6$
(51) $5+4$
(52) $2+8$
(53) $2+9$
(54) $7+1$
(55) $9+9$
(56) $8+7$
(57) $8+6$
(58) $8+9$
(59) $4+7$
(60) $7+2$
(61) $3+4$
(62) $9+3$
(63) $8+2$
(64) $7+6$
(65) $6+1$
(66) $5+3$
(67) $1+9$
(68) $3+3$
(69) $3+10$
(70) $10+10$
(71) $10+7$
(72) $0+8$
(73) $8+4$
(74) $4+4$
(75) $1+1$
(76) $2+10$
(77) $2+2$
(78) $1+6$
(79) $6+7$
(80) $7+3$
(81) $10+10$
(82) $8+1$
(83) $7+10$
(84) $10+9$
(85) $5+2$
(86) $9+5$
(87) $7+5$
(88) $2+7$
(89) $10+2$
(90) $4+1$
(91) $5+9$
(92) $5+6$
(93) $7+8$
(94) $6+9$
(95) $10+8$
(96) $3+5$
(97) $3+6$
(98) $10+4$
(99) $5+5$
(100) $1+4$

Addition I

KEY

No.	Problem	Answer
(1)	9 + 10	19
(2)	8 + 3	11
(3)	3 + 2	5
(4)	6 + 2	8
(5)	1 + 5	6
(6)	0 + 2	2
(7)	4 + 8	12
(8)	3 + 9	12
(9)	2 + 6	8
(10)	0 + 10	10
(11)	8 + 8	16
(12)	9 + 7	16
(13)	10 + 5	15
(14)	6 + 3	9
(15)	9 + 2	11
(16)	8 + 1	9
(17)	1 + 2	3
(18)	0 + 1	1
(19)	3 + 1	4
(20)	6 + 4	10
(21)	4 + 10	14
(22)	9 + 4	13
(23)	5 + 8	13
(24)	5 + 7	12
(25)	4 + 2	6
(26)	6 + 10	16
(27)	2 + 3	5
(28)	2 + 5	7
(29)	2 + 1	3
(30)	3 + 8	11
(31)	4 + 6	10
(32)	7 + 7	14
(33)	6 + 6	12
(34)	8 + 5	13
(35)	3 + 7	10
(36)	1 + 3	4
(37)	0 + 4	4
(38)	2 + 4	6
(39)	7 + 4	11
(40)	6 + 8	14
(41)	1 + 8	9
(42)	10 + 3	13
(43)	9 + 8	17
(44)	5 + 10	15
(45)	10 + 6	16
(46)	7 + 9	16
(47)	6 + 5	11
(48)	4 + 5	9
(49)	4 + 9	13
(50)	9 + 6	15
(51)	5 + 4	9
(52)	2 + 8	10
(53)	2 + 9	11
(54)	7 + 1	8
(55)	9 + 9	18
(56)	8 + 7	15
(57)	8 + 6	14
(58)	8 + 9	17
(59)	4 + 7	11
(60)	7 + 2	9
(61)	3 + 4	7
(62)	9 + 3	12
(63)	8 + 2	10
(64)	7 + 6	13
(65)	6 + 1	7
(66)	5 + 3	8
(67)	1 + 9	10
(68)	3 + 3	6
(69)	3 + 10	13
(70)	10 + 10	20
(71)	10 + 7	17
(72)	0 + 8	8
(73)	8 + 4	12
(74)	4 + 4	8
(75)	1 + 1	2
(76)	2 + 10	12
(77)	2 + 2	4
(78)	1 + 6	7
(79)	6 + 7	13
(80)	7 + 3	10
(81)	10 + 10	20
(82)	8 + 1	9
(83)	7 + 10	17
(84)	10 + 9	19
(85)	5 + 2	7
(86)	9 + 5	14
(87)	7 + 5	12
(88)	2 + 7	9
(89)	10 + 2	12
(90)	4 + 1	5
(91)	5 + 9	14
(92)	5 + 6	11
(93)	7 + 8	15
(94)	6 + 9	15
(95)	10 + 8	18
(96)	3 + 5	8
(97)	3 + 6	9
(98)	10 + 4	14
(99)	5 + 5	10
(100)	1 + 4	5

Addition II

Name ______________________ Date ______________ Score ______ Time ______

(1) 0 + 1 ☐
(2) 6 + 2 ☐
(3) 9 + 4 ☐
(4) 3 + 9 ☐
(5) 6 + 10 ☐
(6) 9 + 7 ☐
(7) 3 + 8 ☐
(8) 8 + 1 ☐
(9) 8 + 5 ☐
(10) 6 + 4 ☐
(11) 2 + 4 ☐
(12) 5 + 7 ☐
(13) 10 + 3 ☐
(14) 2 + 5 ☐
(15) 7 + 9 ☐
(16) 7 + 7 ☐
(17) 9 + 6 ☐
(18) 1 + 3 ☐
(19) 7 + 1 ☐
(20) 6 + 8 ☐
(21) 8 + 9 ☐
(22) 5 + 10 ☐
(23) 9 + 3 ☐
(24) 4 + 5 ☐
(25) 5 + 3 ☐
(26) 1 + 2 ☐
(27) 0 + 5 ☐
(28) 4 + 10 ☐
(29) 3 + 1 ☐
(30) 4 + 2 ☐
(31) 0 + 7 ☐
(32) 4 + 3 ☐
(33) 9 + 1 ☐
(34) 6 + 6 ☐
(35) 5 + 8 ☐
(36) 0 + 4 ☐
(37) 2 + 3 ☐
(38) 0 + 3 ☐
(39) 1 + 8 ☐
(40) 4 + 6 ☐
(41) 10 + 6 ☐
(42) 3 + 7 ☐
(43) 4 + 9 ☐
(44) 7 + 4 ☐
(45) 2 + 9 ☐
(46) 9 + 8 ☐
(47) 8 + 6 ☐
(48) 6 + 5 ☐
(49) 3 + 4 ☐
(50) 5 + 4 ☐
(51) 6 + 3 ☐
(52) 5 + 5 ☐
(53) 0 + 10 ☐
(54) 10 + 8 ☐
(55) 0 + 2 ☐
(56) 5 + 9 ☐
(57) 8 + 3 ☐
(58) 7 + 5 ☐
(59) 3 + 6 ☐
(60) 7 + 10 ☐
(61) 7 + 8 ☐
(62) 6 + 7 ☐
(63) 10 + 2 ☐
(64) 1 + 1 ☐
(65) 5 + 2 ☐
(66) 10 + 7 ☐
(67) 8 + 10 ☐
(68) 1 + 9 ☐
(69) 2 + 2 ☐
(70) 8 + 2 ☐
(71) 8 + 4 ☐
(72) 4 + 7 ☐
(73) 3 + 10 ☐
(74) 9 + 9 ☐
(75) 6 + 1 ☐
(76) 9 + 2 ☐
(77) 3 + 5 ☐
(78) 1 + 10 ☐
(79) 5 + 6 ☐
(80) 8 + 8 ☐
(81) 2 + 7 ☐
(82) 4 + 8 ☐
(83) 10 + 9 ☐
(84) 3 + 2 ☐
(85) 7 + 3 ☐
(86) 10 + 4 ☐
(87) 2 + 10 ☐
(88) 6 + 9 ☐
(89) 0 + 8 ☐
(90) 4 + 1 ☐
(91) 3 + 3 ☐
(92) 9 + 5 ☐
(93) 7 + 6 ☐
(94) 10 + 1 ☐
(95) 7 + 2 ☐
(96) 1 + 6 ☐
(97) 8 + 7 ☐
(98) 4 + 4 ☐
(99) 2 + 8 ☐
(100) 10 + 10 ☐

Addition II

KEY

(1)	0 + 1	1	(26)	1 + 2	3	(51)	6 + 3	9	(76)	9 + 2	11
(2)	6 + 2	8	(27)	0 + 5	5	(52)	5 + 5	10	(77)	3 + 5	8
(3)	9 + 4	13	(28)	4 + 10	14	(53)	0 + 10	10	(78)	1 + 10	11
(4)	3 + 9	12	(29)	3 + 1	4	(54)	10 + 8	18	(79)	5 + 6	11
(5)	6 + 10	16	(30)	4 + 2	6	(55)	0 + 2	2	(80)	8 + 8	16
(6)	9 + 7	16	(31)	0 + 7	7	(56)	5 + 9	14	(81)	2 + 7	9
(7)	3 + 8	11	(32)	4 + 3	7	(57)	8 + 3	11	(82)	4 + 8	12
(8)	8 + 1	9	(33)	9 + 1	10	(58)	7 + 5	12	(83)	10 + 9	19
(9)	8 + 5	13	(34)	6 + 6	12	(59)	3 + 6	9	(84)	3 + 2	5
(10)	6 + 4	10	(35)	5 + 8	13	(60)	7 + 10	17	(85)	7 + 3	10
(11)	2 + 4	6	(36)	0 + 4	4	(61)	7 + 8	15	(86)	10 + 4	14
(12)	5 + 7	12	(37)	2 + 3	5	(62)	6 + 7	13	(87)	2 + 10	12
(13)	10 + 3	13	(38)	0 + 3	3	(63)	10 + 2	12	(88)	6 + 9	15
(14)	2 + 5	7	(39)	1 + 8	9	(64)	1 + 1	2	(89)	0 + 8	8
(15)	7 + 9	16	(40)	4 + 6	10	(65)	5 + 2	7	(90)	4 + 1	5
(16)	7 + 7	14	(41)	10 + 6	16	(66)	10 + 7	17	(91)	3 + 3	6
(17)	9 + 6	15	(42)	3 + 7	10	(67)	8 + 10	18	(92)	9 + 5	14
(18)	1 + 3	4	(43)	4 + 9	13	(68)	1 + 9	10	(93)	7 + 6	13
(19)	7 + 1	8	(44)	7 + 4	11	(69)	2 + 2	4	(94)	10 + 1	11
(20)	6 + 8	14	(45)	2 + 9	11	(70)	8 + 2	10	(95)	7 + 2	9
(21)	8 + 9	17	(46)	9 + 8	17	(71)	8 + 4	12	(96)	1 + 6	7
(22)	5 + 10	15	(47)	8 + 6	14	(72)	4 + 7	11	(97)	8 + 7	15
(23)	9 + 3	12	(48)	6 + 5	11	(73)	3 + 10	13	(98)	4 + 4	8
(24)	4 + 5	9	(49)	3 + 4	7	(74)	9 + 9	18	(99)	2 + 8	10
(25)	5 + 3	8	(50)	5 + 4	9	(75)	6 + 1	7	(100)	10 + 10	20

Addition III

Name ______________________ Date ____________ Score ______ Time ______

(1)	9 + 10	(26)	3 + 8	(51)	4 + 7	(76)	2 + 7
(2)	0 + 2	(27)	3 + 7	(52)	7 + 6	(77)	2 + 2
(3)	8 + 8	(28)	6 + 8	(53)	2 + 9	(78)	10 + 1
(4)	8 + 1	(29)	2 + 1	(54)	8 + 9	(79)	7 + 5
(5)	1 + 5	(30)	8 + 5	(55)	8 + 2	(80)	5 + 6
(6)	0 + 10	(31)	7 + 4	(56)	3 + 3	(81)	8 + 10
(7)	9 + 2	(32)	5 + 10	(57)	8 + 6	(82)	9 + 5
(8)	6 + 4	(33)	6 + 6	(58)	9 + 3	(83)	3 + 5
(9)	2 + 6	(34)	2 + 4	(59)	1 + 9	(84)	5 + 9
(10)	6 + 3	(35)	9 + 8	(60)	0 + 8	(85)	5 + 2
(11)	3 + 1	(36)	4 + 5	(61)	3 + 4	(86)	4 + 1
(12)	5 + 7	(37)	0 + 4	(62)	5 + 3	(87)	10 + 8
(13)	10 + 5	(38)	10 + 3	(63)	10 + 7	(88)	1 + 4
(14)	0 + 1	(39)	6 + 5	(64)	2 + 10	(89)	10 + 2
(15)	5 + 8	(40)	2 + 8	(65)	6 + 1	(90)	6 + 9
(16)	2 + 5	(41)	1 + 8	(66)	10 + 10	(91)	5 + 5
(17)	1 + 2	(42)	7 + 9	(67)	1 + 1	(92)	6 + 2
(18)	9 + 4	(43)	5 + 4	(68)	7 + 3	(93)	7 + 8
(19)	2 + 3	(44)	8 + 7	(69)	3 + 10	(94)	10 + 4
(20)	7 + 7	(45)	10 + 6	(70)	4 + 4	(95)	3 + 2
(21)	4 + 10	(46)	9 + 6	(71)	6 + 7	(96)	3 + 9
(22)	6 + 10	(47)	9 + 9	(72)	10 + 9	(97)	3 + 6
(23)	4 + 6	(48)	7 + 2	(73)	8 + 4	(98)	8 + 3
(24)	1 + 3	(49)	4 + 9	(74)	1 + 6	(99)	4 + 8
(25)	4 + 2	(50)	7 + 1	(75)	7 + 10	(100)	9 + 7

Addition III

KEY

No.	Problem	Answer	No.	Problem	Answer	No.	Problem	Answer	No.	Problem	Answer
(1)	9 + 10	19	(26)	3 + 8	11	(51)	4 + 7	11	(76)	2 + 7	9
(2)	0 + 2	2	(27)	3 + 7	10	(52)	7 + 6	13	(77)	2 + 2	4
(3)	8 + 8	16	(28)	6 + 8	14	(53)	2 + 9	11	(78)	10 + 1	11
(4)	8 + 1	9	(29)	2 + 1	3	(54)	8 + 9	17	(79)	7 + 5	12
(5)	1 + 5	6	(30)	8 + 5	13	(55)	8 + 2	10	(80)	5 + 6	11
(6)	0 + 10	10	(31)	7 + 4	11	(56)	3 + 3	6	(81)	8 + 10	18
(7)	9 + 2	11	(32)	5 + 10	15	(57)	8 + 6	14	(82)	9 + 5	14
(8)	6 + 4	10	(33)	6 + 6	12	(58)	9 + 3	12	(83)	3 + 5	8
(9)	2 + 6	8	(34)	2 + 4	6	(59)	1 + 9	10	(84)	5 + 9	14
(10)	6 + 3	9	(35)	9 + 8	17	(60)	0 + 8	8	(85)	5 + 2	7
(11)	3 + 1	4	(36)	4 + 5	9	(61)	3 + 4	7	(86)	4 + 1	5
(12)	5 + 7	12	(37)	0 + 4	4	(62)	5 + 3	8	(87)	10 + 8	18
(13)	10 + 5	15	(38)	10 + 3	13	(63)	10 + 7	17	(88)	1 + 4	5
(14)	0 + 1	1	(39)	6 + 5	11	(64)	2 + 10	12	(89)	10 + 2	12
(15)	5 + 8	13	(40)	2 + 8	10	(65)	6 + 1	7	(90)	6 + 9	15
(16)	2 + 5	7	(41)	1 + 8	9	(66)	10 + 10	20	(91)	5 + 5	10
(17)	1 + 2	3	(42)	7 + 9	16	(67)	1 + 1	2	(92)	6 + 2	8
(18)	9 + 4	13	(43)	5 + 4	9	(68)	7 + 3	10	(93)	7 + 8	15
(19)	2 + 3	5	(44)	8 + 7	15	(69)	3 + 10	13	(94)	10 + 4	14
(20)	7 + 7	14	(45)	10 + 6	16	(70)	4 + 4	8	(95)	3 + 2	5
(21)	4 + 10	14	(46)	9 + 6	15	(71)	6 + 7	13	(96)	3 + 9	12
(22)	6 + 10	16	(47)	9 + 9	18	(72)	10 + 9	19	(97)	3 + 6	9
(23)	4 + 6	10	(48)	7 + 2	9	(73)	8 + 4	12	(98)	8 + 3	11
(24)	1 + 3	4	(49)	4 + 9	13	(74)	1 + 6	7	(99)	4 + 8	12
(25)	4 + 2	6	(50)	7 + 1	8	(75)	7 + 10	17	(100)	9 + 7	16

Addition IV

Name ____________________________ Date ______________ Score ______ Time ______

(1) 9 + 6
(2) 4 + 9
(3) 4 + 5
(4) 6 + 5
(5) 7 + 9
(6) 10 + 6
(7) 5 + 10
(8) 9 + 8
(9) 10 + 3
(10) 1 + 8
(11) 6 + 8
(12) 7 + 4
(13) 2 + 4
(14) 0 + 4
(15) 1 + 3
(16) 3 + 7
(17) 8 + 5
(18) 6 + 6
(19) 7 + 7
(20) 4 + 6
(21) 3 + 8
(22) 2 + 1
(23) 2 + 5
(24) 2 + 3
(25) 6 + 10

(26) 5 + 4
(27) 2 + 9
(28) 9 + 9
(29) 8 + 6
(30) 4 + 7
(31) 3 + 4
(32) 8 + 2
(33) 6 + 1
(34) 1 + 9
(35) 3 + 10
(36) 10 + 7
(37) 8 + 4
(38) 1 + 1
(39) 5 + 5
(40) 3 + 6
(41) 10 + 8
(42) 7 + 8
(43) 5 + 9
(44) 10 + 2
(45) 7 + 5
(46) 5 + 2
(47) 7 + 10
(48) 8 + 10
(49) 6 + 7
(50) 2 + 2

(51) 4 + 2
(52) 5 + 8
(53) 4 + 10
(54) 3 + 1
(55) 1 + 2
(56) 9 + 2
(57) 10 + 5
(58) 8 + 8
(59) 2 + 6
(60) 4 + 8
(61) 1 + 5
(62) 3 + 2
(63) 9 + 10
(64) 2 + 8
(65) 7 + 1
(66) 8 + 7
(67) 8 + 9
(68) 7 + 2
(69) 9 + 3
(70) 7 + 6
(71) 5 + 3
(72) 3 + 3
(73) 10 + 10
(74) 0 + 8
(75) 4 + 4

(76) 2 + 10
(77) 1 + 6
(78) 7 + 3
(79) 10 + 1
(80) 10 + 9
(81) 9 + 5
(82) 2 + 7
(83) 4 + 1
(84) 5 + 6
(85) 6 + 9
(86) 3 + 5
(87) 10 + 4
(88) 1 + 4
(89) 8 + 3
(90) 6 + 2
(91) 0 + 2
(92) 3 + 9
(93) 0 + 10
(94) 9 + 7
(95) 6 + 3
(96) 8 + 1
(97) 0 + 1
(98) 6 + 4
(99) 9 + 4
(100) 5 + 7

Addition IV

KEY

No.	Problem	Answer
(1)	9 + 6	15
(2)	4 + 9	13
(3)	4 + 5	9
(4)	6 + 5	11
(5)	7 + 9	16
(6)	10 + 6	16
(7)	5 + 10	15
(8)	9 + 8	17
(9)	10 + 3	13
(10)	1 + 8	9
(11)	6 + 8	14
(12)	7 + 4	11
(13)	2 + 4	6
(14)	0 + 4	4
(15)	1 + 3	4
(16)	3 + 7	10
(17)	8 + 5	13
(18)	6 + 6	12
(19)	7 + 7	14
(20)	4 + 6	10
(21)	3 + 8	11
(22)	2 + 1	3
(23)	2 + 5	7
(24)	2 + 3	5
(25)	6 + 10	16
(26)	5 + 4	9
(27)	2 + 9	11
(28)	9 + 9	18
(29)	8 + 6	14
(30)	4 + 7	11
(31)	3 + 4	7
(32)	8 + 2	10
(33)	6 + 1	7
(34)	1 + 9	10
(35)	3 + 10	13
(36)	10 + 7	17
(37)	8 + 4	12
(38)	1 + 1	2
(39)	5 + 5	10
(40)	3 + 6	9
(41)	10 + 8	18
(42)	7 + 8	15
(43)	5 + 9	14
(44)	10 + 2	12
(45)	7 + 5	12
(46)	5 + 2	7
(47)	7 + 10	17
(48)	8 + 10	18
(49)	6 + 7	13
(50)	2 + 2	4
(51)	4 + 2	6
(52)	5 + 8	13
(53)	4 + 10	14
(54)	3 + 1	4
(55)	1 + 2	3
(56)	9 + 2	11
(57)	10 + 5	15
(58)	8 + 8	16
(59)	2 + 6	8
(60)	4 + 8	12
(61)	1 + 5	6
(62)	3 + 2	5
(63)	9 + 10	19
(64)	2 + 8	10
(65)	7 + 1	8
(66)	8 + 7	15
(67)	8 + 9	17
(68)	7 + 2	9
(69)	9 + 3	12
(70)	7 + 6	13
(71)	5 + 3	8
(72)	3 + 3	6
(73)	10 + 10	20
(74)	0 + 8	8
(75)	4 + 4	8
(76)	2 + 10	12
(77)	1 + 6	7
(78)	7 + 3	10
(79)	10 + 1	11
(80)	10 + 9	19
(81)	9 + 5	14
(82)	2 + 7	9
(83)	4 + 1	5
(84)	5 + 6	11
(85)	6 + 9	15
(86)	3 + 5	8
(87)	10 + 4	14
(88)	1 + 4	5
(89)	8 + 3	11
(90)	6 + 2	8
(91)	0 + 2	2
(92)	3 + 9	12
(93)	0 + 10	10
(94)	9 + 7	16
(95)	6 + 3	9
(96)	8 + 1	9
(97)	0 + 1	1
(98)	6 + 4	10
(99)	9 + 4	13
(100)	5 + 7	12

Addition V

Name ____________________ Date ____________ Score ______ Time ______

(1) 1 + 1
(2) 4 + 2
(3) 9 + 6
(4) 1 + 4
(5) 9 + 10
(6) 5 + 4
(7) 6 + 10
(8) 2 + 10
(9) 5 + 7
(10) 4 + 9
(11) 4 + 4
(12) 5 + 5
(13) 10 + 4
(14) 8 + 4
(15) 4 + 5
(16) 5 + 8
(17) 9 + 4
(18) 6 + 5
(19) 0 + 8
(20) 3 + 6
(21) 3 + 5
(22) 10 + 7
(23) 7 + 9
(24) 4 + 10
(25) 6 + 4
(26) 10 + 6
(27) 10 + 10
(28) 6 + 9
(29) 10 + 8
(30) 3 + 10
(31) 3 + 1
(32) 5 + 10
(33) 0 + 1
(34) 9 + 8
(35) 3 + 3
(36) 7 + 8
(37) 6 + 5
(38) 1 + 9
(39) 10 + 3
(40) 1 + 2
(41) 8 + 1
(42) 5 + 3
(43) 1 + 8
(44) 5 + 9
(45) 4 + 1
(46) 6 + 8
(47) 6 + 1
(48) 9 + 2
(49) 6 + 3
(50) 7 + 4
(51) 7 + 6
(52) 10 + 2
(53) 2 + 7
(54) 8 + 2
(55) 2 + 4
(56) 10 + 5
(57) 9 + 7
(58) 0 + 4
(59) 9 + 3
(60) 7 + 5
(61) 9 + 5
(62) 3 + 4
(63) 1 + 3
(64) 8 + 8
(65) 0 + 10
(66) 3 + 7
(67) 7 + 2
(68) 5 + 2
(69) 10 + 9
(70) 4 + 7
(71) 8 + 5
(72) 2 + 6
(73) 3 + 9
(74) 6 + 6
(75) 8 + 9
(76) 7 + 10
(77) 10 + 1
(78) 8 + 6
(79) 7 + 7
(80) 4 + 8
(81) 0 + 2
(82) 4 + 6
(83) 8 + 7
(84) 8 + 10
(85) 7 + 3
(86) 9 + 9
(87) 3 + 8
(88) 1 + 5
(89) 6 + 2
(90) 2 + 1
(91) 7 + 1
(92) 6 + 7
(93) 1 + 6
(94) 2 + 9
(95) 2 + 5
(96) 3 + 4
(97) 8 + 3
(98) 2 + 3
(99) 2 + 8
(100) 2 + 2

Addition V

KEY

No.	Problem	Answer
(1)	$1 + 1$	2
(2)	$4 + 2$	6
(3)	$9 + 6$	15
(4)	$1 + 4$	5
(5)	$9 + 10$	19
(6)	$5 + 4$	9
(7)	$6 + 10$	16
(8)	$2 + 10$	12
(9)	$5 + 7$	12
(10)	$4 + 9$	13
(11)	$4 + 4$	8
(12)	$5 + 5$	10
(13)	$10 + 4$	14
(14)	$8 + 4$	12
(15)	$4 + 5$	9
(16)	$5 + 8$	13
(17)	$9 + 4$	13
(18)	$6 + 5$	11
(19)	$0 + 8$	8
(20)	$3 + 6$	9
(21)	$3 + 5$	8
(22)	$10 + 7$	17
(23)	$7 + 9$	16
(24)	$4 + 10$	14
(25)	$6 + 4$	10
(26)	$10 + 6$	16
(27)	$10 + 10$	20
(28)	$6 + 9$	15
(29)	$10 + 8$	18
(30)	$3 + 10$	13
(31)	$3 + 1$	4
(32)	$5 + 10$	15
(33)	$0 + 1$	1
(34)	$9 + 8$	17
(35)	$3 + 3$	6
(36)	$7 + 8$	15
(37)	$6 + 5$	11
(38)	$1 + 9$	10
(39)	$10 + 3$	13
(40)	$1 + 2$	3
(41)	$8 + 1$	9
(42)	$5 + 3$	8
(43)	$1 + 8$	9
(44)	$5 + 9$	14
(45)	$4 + 1$	5
(46)	$6 + 8$	14
(47)	$6 + 1$	7
(48)	$9 + 2$	11
(49)	$6 + 3$	9
(50)	$7 + 4$	11
(51)	$7 + 6$	13
(52)	$10 + 2$	12
(53)	$2 + 7$	9
(54)	$8 + 2$	10
(55)	$2 + 4$	6
(56)	$10 + 5$	15
(57)	$9 + 7$	16
(58)	$0 + 4$	4
(59)	$9 + 3$	12
(60)	$7 + 5$	12
(61)	$9 + 5$	14
(62)	$3 + 4$	7
(63)	$1 + 3$	4
(64)	$8 + 8$	16
(65)	$0 + 10$	10
(66)	$3 + 7$	10
(67)	$7 + 2$	9
(68)	$5 + 2$	7
(69)	$10 + 9$	19
(70)	$4 + 7$	11
(71)	$8 + 5$	13
(72)	$2 + 6$	8
(73)	$3 + 9$	12
(74)	$6 + 6$	12
(75)	$8 + 9$	17
(76)	$7 + 10$	17
(77)	$10 + 1$	11
(78)	$8 + 6$	14
(79)	$7 + 7$	14
(80)	$4 + 8$	12
(81)	$0 + 2$	2
(82)	$4 + 6$	10
(83)	$8 + 7$	15
(84)	$8 + 10$	18
(85)	$7 + 3$	10
(86)	$9 + 9$	18
(87)	$3 + 8$	11
(88)	$1 + 5$	6
(89)	$6 + 2$	8
(90)	$2 + 1$	3
(91)	$7 + 1$	8
(92)	$6 + 7$	13
(93)	$1 + 6$	7
(94)	$2 + 9$	11
(95)	$2 + 5$	7
(96)	$3 + 4$	7
(97)	$8 + 3$	11
(98)	$2 + 3$	5
(99)	$2 + 8$	10
(100)	$2 + 2$	4

Folders and Records

Folders and Records

This section contains reproducible pages and forms to help you and your students keep track of progress toward mastering addition facts.

Student Folders

Each student in my class has a folder. It can be a plain manila folder or a pocket portfolio. My students have pocket portfolios color-coded for each major subject. Staple or tape any (or all) of the following into the folder. Tailor it to your own preferences!

Addition Stickers, p. 54. I staple this inside the folder on the left. Tests which are completed in three minutes or less are graded by the teacher. If a child gets 100% correct, note the date the test was passed on this sheet and award a sticker. Adjust the mastery standard for young students if necessary.

Record of Timed Tests, p. 55. I duplicate four of these sheets, one for each quarter, and note the days and dates school will be in session (no weekends or holidays) under "Date." For example: T 9/5/95, W 9/6/95, Th 9/7/95, F 9/8/95, etc. I use these as masters and copy a set for each student on colored paper. They are stapled into the folders on the right. After each daily test, each child notes the test she took: 0-1, 2, 3, 0-3, etc. or a ditto symbol if the same test was taken as on the previous day. The child then notes her score and time. If a child is absent, the note ***Absent*** can be written in. After results are recorded, the test papers can be recycled or discarded.

Graph, p. 56. Teachers may prefer to have students graph their scores rather than keep track with the Record of Timed Tests. Teachers should write in dates before reproducing. Different colored markers can be used for different tests.

Addition Chart, p. 53 or p. 85. A chart of addition facts can be attached to the folder. This relieves pressure on students, especially those for whom learning the facts is difficult, and especially when they're first beginning with a new timed test. Alternately, you may wish to post a large chart on the wall (in the back of the room so it's not *too* easy to see). Because of the repetition built into the tests, students won't need to refer to the chart for long. Checking a chart takes time. It is nearly impossible to pass in three minutes or less if it's used, or if students count on their fingers.

Answer Keys, pp. 8-50. Once students begin progressing at their own speeds through the timed test worksheets, they will need answer keys so they can check their own papers on days when they do not finish in less than three minutes. I duplicate all addition keys for each child two-sided on yellow paper and staple them together to be kept in the folders. I collect them for reuse at the end of the school year.

Flash Cards, pp. 61-82. Flash cards in plastic sandwich bags may also be stored in the folders until they're sent home.

Teacher Record-Keeping

Time Chart. A large chart like the example below can be posted in the front of the room. The teacher (or an aide) points to each time as it elapses after a timed test begins. When a child completes the timed test, he/she looks up and writes the time to which the teacher is pointing.

1:00	2:00	3:00	4:00
1:05	2:05	3:05	4:05
1:10	2:10	3:10	4:10
1:15	2:15	3:15	4:15
1:20	2:20	3:20	4:20
1:25	2:25	3:25	4:25
1:30	2:30	3:30	4:30
1:35	2:35	3:35	4:35
1:40	2:40	3:40	4:40
1:45	2:45	3:45	4:45
1:50	2:50	3:50	4:50
1:55	2:55	3:55	4:55
			5:00

Class Record, p. 57. This chart can be duplicated or a gradebook can be used. List student names on the left. When a child passes a test, check the box or make a small note of the date the test was passed. An alternative is to use a large wall chart. Children may write in the dates they pass!

Congratulations Certificates, p. 58. In place of or in addition to the sticker in the folder, you can fill out a certificate for a child to take home and decorate *it* with a sticker.

Addition Chart

The addition chart on the right may be reproduced and glued into the folders behind the Record of Timed Tests, where it is available but not *easily* available. Teach students how to use it. The answers are where the addends intersect. See p. 52 for alternatives and further information about the addition chart.

	0	**1**	**2**	**3**	**4**	**5**	**6**	**7**	**8**	**9**	**10**
0	0	1	2	3	4	5	6	7	8	9	10
1	1	2	3	4	5	6	7	8	9	10	11
2	2	3	4	5	6	7	8	9	10	11	12
3	3	4	5	6	7	8	9	10	11	12	13
4	4	5	6	7	8	9	10	11	12	13	14
5	5	6	7	8	9	10	11	12	13	14	15
6	6	7	8	9	10	11	12	13	14	15	16
7	7	8	9	10	11	12	13	14	15	16	17
8	8	9	10	11	12	13	14	15	16	17	18
9	9	10	11	12	13	14	15	16	17	18	19
10	10	11	12	13	14	15	16	17	18	19	20

Addition Stickers

0-1	Date passed ________
2	Date passed ________
3	Date passed ________
0-3	Date passed ________
4	Date passed ________
0-4	Date passed ________
5	Date passed ________
0-5	Date passed ________
6	Date passed ________

0-6	Date passed ________
7	Date passed ________
0-7	Date passed ________
8	Date passed ________
0-8	Date passed ________
9	Date passed ________
0-9	Date passed ________
10	Date passed ________
0-10	Date passed ________

Record of Timed Tests

Name____

DATE	TEST	SCORE	TIME

DATE	TEST	SCORE	TIME

Name ______________________________

Test ______ ______ ______ ______ ______ ______ ______ ______ ______ ______ ______ ______ ______ ______ ______

Time ______ ______ ______ ______ ______ ______ ______ ______ ______ ______ ______ ______ ______ ______ ______

100
95
90
85
80
75
70
65
60
55
50
45
40
35
30
25
20
15
10
5

Dates

Addition	0-1	2	3	0-3	4	0-4	5	0-5	6	0-6	7	0-7	8	0-8	9	0-9	10	ALL

CONGRATULATIONS!

______________________ completed 100 addition facts correctly in three minutes or less today and passed the ______ test! This is an accomplishment to *celebrate!*

____
Teacher

____
Date

CONGRATULATIONS!

______________________ completed 100 addition facts correctly in three minutes or less today and passed the ______ test! This is an accomplishment to *celebrate!*

____
Teacher

____
Date

CONGRATULATIONS!

______________________ completed 100 addition facts correctly in three minutes or less today and passed the ______ test! This is an accomplishment to *celebrate!*

____
Teacher

____
Date

CONGRATULATIONS!

______________________ completed 100 addition facts correctly in three minutes or less today and passed the ______ test! This is an accomplishment to *celebrate!*

____
Teacher

____
Date

Flash Card Masters

Flash Cards

On the following pages are black-line masters for personal flash cards. Here are some ideas and possibilities for using them:

1) Flash cards can be run two-sided on tagboard or heavy paper.

2) Students can practice a set of facts with flash cards alone or with a friend before attempting the timed tests.

3) Students can store personal flash cards in plastic sandwich bags in their math folders or in boxes in their desks.

4) Flash cards can be run one-sided and used for games such as Bingo (see pp. 84, 87-88). Students can create their own games for each other to play.

5) Flash cards can be sent home with students for practice with parents.

$0 + 0 =$	$0 + 1 =$	$0 + 2 =$
$0 + 3 =$	$0 + 4 =$	$0 + 5 =$
$0 + 6 =$	$0 + 7 =$	$0 + 8 =$
$0 + 9 =$	$0 + 10 =$	ZEROS

0 + 0 = 0
0 + 1 = 1
0 + 2 = 2
0 + 3 = 3
0 + 4 = 4
0 + 5 = 5
0 + 6 = 6
0 + 7 = 7
0 + 8 = 8
0 + 9 = 9
0 + 10 = 10

0 + 2 = 2

0 + 1 = 1

0 + 0 = 0

0 + 5 = 5

0 + 4 = 4

0 + 3 = 3

0 + 8 = 8

0 + 7 = 7

0 + 6 = 6

0 + 10 = 10

0 + 9 = 9

$1 + 0 =$	$1 + 1 =$	$1 + 2 =$
$1 + 3 =$	$1 + 4 =$	$1 + 5 =$
$1 + 6 =$	$1 + 7 =$	$1 + 8 =$
$1 + 9 =$	$1 + 10 =$	ONES

1 + 2 = 3 | 1 + 1 = 2 | 1 + 0 = 1

1 + 5 = 6 | 1 + 4 = 5 | 1 + 3 = 4

1 + 8 = 9 | 1 + 7 = 8 | 1 + 6 = 7

1 + 10 = 11 | 1 + 9 = 10

1 + 0 = 1
1 + 1 = 2
1 + 2 = 3
1 + 3 = 4
1 + 4 = 5
1 + 5 = 6
1 + 6 = 7
1 + 7 = 8
1 + 8 = 9
1 + 9 = 10
1 + 10 = 11

$2 + 0 =$	$2 + 1 =$	$2 + 2 =$
$2 + 3 =$	$2 + 4 =$	$2 + 5 =$
$2 + 6 =$	$2 + 7 =$	$2 + 8 =$
$2 + 9 =$	$2 + 10 =$	TWOS

$2 + 0 = 2$
$2 + 1 = 3$
$2 + 2 = 4$
$2 + 3 = 5$
$2 + 4 = 6$
$2 + 5 = 7$
$2 + 6 = 8$
$2 + 7 = 9$
$2 + 8 = 10$
$2 + 9 = 11$
$2 + 10 = 12$

$2 + 2 = 4$

$2 + 1 = 3$

$2 + 0 = 2$

$2 + 5 = 7$

$2 + 4 = 6$

$2 + 3 = 5$

$2 + 8 = 10$

$2 + 7 = 9$

$2 + 6 = 8$

$2 + 10 = 12$

$2 + 9 = 11$

$3+0=$	$3+1=$	$3+2=$
$3+3=$	$3+4=$	$3+5=$
$3+6=$	$3+7=$	$3+8=$
$3+9=$	$3+10=$	THREES

3 + 0 = 3
3 + 1 = 4
3 + 2 = 5
3 + 3 = 6
3 + 4 = 7
3 + 5 = 8
3 + 6 = 9
3 + 7 = 10
3 + 8 = 11
3 + 9 = 12
3 + 10 = 13

3 + 8 = 11

3 + 5 = 8

3 + 2 = 5

3 + 10 = 13

3 + 7 = 10

3 + 4 = 7

3 + 1 = 4

3 + 9 = 12

3 + 6 = 9

3 + 3 = 6

3 + 0 = 3

$4 + 0 =$	$4 + 1 =$	$4 + 2 =$
$4 + 3 =$	$4 + 4 =$	$4 + 5 =$
$4 + 6 =$	$4 + 7 =$	$4 + 8 =$
$4 + 9 =$	$4 + 10 =$	FOURS

$4 + 0 = 4$
$4 + 1 = 5$
$4 + 2 = 6$
$4 + 3 = 7$
$4 + 4 = 8$
$4 + 5 = 9$
$4 + 6 = 10$
$4 + 7 = 11$
$4 + 8 = 12$
$4 + 9 = 13$
$4 + 10 = 14$

$4 + 2 = 6$

$4 + 1 = 5$

$4 + 0 = 4$

$4 + 5 = 9$

$4 + 4 = 8$

$4 + 3 = 7$

$4 + 8 = 12$

$4 + 7 = 11$

$4 + 6 = 10$

$4 + 10 = 14$

$4 + 9 = 13$

$5 + 0 =$	$5 + 1 =$	$5 + 2 =$
$5 + 3 =$	$5 + 4 =$	$5 + 5 =$
$5 + 6 =$	$5 + 7 =$	$5 + 8 =$
$5 + 9 =$	$5 + 10 =$	FIVES

5 + 0 = 5
5 + 1 = 6
5 + 2 = 7
5 + 3 = 8
5 + 4 = 9
5 + 5 = 10
5 + 6 = 11
5 + 7 = 12
5 + 8 = 13
5 + 9 = 14
5 + 10 = 15

5 + 2 = 7

5 + 1 = 6

5 + 0 = 5

5 + 5 = 10

5 + 4 = 9

5 + 3 = 8

5 + 8 = 13

5 + 7 = 12

5 + 6 = 11

5 + 10 = 15

5 + 9 = 14

$6 + 0 =$	$6 + 1 =$	$6 + 2 =$
$6 + 3 =$	$6 + 4 =$	$6 + 5 =$
$6 + 6 =$	$6 + 7 =$	$6 + 8 =$
$6 + 9 =$	$6 + 10 =$	SIXES

$6 + 2 = 8$ $6 + 1 = 7$ $6 + 0 = 6$

$6 + 5 = 11$ $6 + 4 = 10$ $6 + 3 = 9$

$6 + 8 = 14$ $6 + 7 = 13$ $6 + 6 = 12$

$6 + 10 = 16$ $6 + 9 = 15$

$6 + 0 = 6$
$6 + 1 = 7$
$6 + 2 = 8$
$6 + 3 = 9$
$6 + 4 = 10$
$6 + 5 = 11$
$6 + 6 = 12$
$6 + 7 = 13$
$6 + 8 = 14$
$6 + 9 = 15$
$6 + 10 = 16$

$7 + 0 =$	$7 + 1 =$	$7 + 2 =$
$7 + 3 =$	$7 + 4 =$	$7 + 5 =$
$7 + 6 =$	$7 + 7 =$	$7 + 8 =$
$7 + 9 =$	$7 + 10 =$	SEVENS

$7 + 0 = 7$
$7 + 1 = 8$
$7 + 2 = 9$
$7 + 3 = 10$
$7 + 4 = 11$
$7 + 5 = 12$
$7 + 6 = 13$
$7 + 7 = 14$
$7 + 8 = 15$
$7 + 9 = 16$
$7 + 10 = 17$

$7 + 2 = 9$

$7 + 1 = 8$

$7 + 0 = 7$

$7 + 5 = 12$

$7 + 4 = 11$

$7 + 3 = 10$

$7 + 8 = 15$

$7 + 7 = 14$

$7 + 6 = 13$

$7 + 10 = 17$

$7 + 9 = 16$

$8 + 0 =$	$8 + 1 =$	$8 + 2 =$
$8 + 3 =$	$8 + 4 =$	$8 + 5 =$
$8 + 6 =$	$8 + 7 =$	$8 + 8 =$
$8 + 9 =$	$8 + 10 =$	EIGHTS

8 + 0 = 8
8 + 1 = 9
8 + 2 = 10
8 + 3 = 11
8 + 4 = 12
8 + 5 = 13
8 + 6 = 14
8 + 7 = 15
8 + 8 = 16
8 + 9 = 17
8 + 10 = 18

8 + 8 = 16

8 + 5 = 13

8 + 2 = 10

8 + 10 = 18

8 + 7 = 15

8 + 4 = 12

8 + 1 = 9

8 + 9 = 17

8 + 6 = 14

8 + 3 = 11

8 + 0 = 8

$9 + 0 =$	$9 + 1 =$	$9 + 2 =$
$9 + 3 =$	$9 + 4 =$	$9 + 5 =$
$9 + 6 =$	$9 + 7 =$	$9 + 8 =$
$9 + 9 =$	$9 + 10 =$	NINES

9 + 0 = 9
9 + 1 = 10
9 + 2 = 11
9 + 3 = 12
9 + 4 = 13
9 + 5 = 14
9 + 6 = 15
9 + 7 = 16
9 + 8 = 17
9 + 9 = 18
9 + 10 = 19

9 + 8 = 17

9 + 5 = 14

9 + 2 = 11

9 + 10 = 19

9 + 7 = 16

9 + 4 = 13

9 + 1 = 10

9 + 9 = 18

9 + 6 = 15

9 + 3 = 12

9 + 0 = 9

10 + 0 =	10 + 1 =	10 + 2 =
10 + 3 =	10 + 4 =	10 + 5 =
10 + 6 =	10 + 7 =	10 + 8 =
10 + 9 =	10 + 10 =	TENS

10 + 0 = 10
10 + 1 = 11
10 + 2 = 12
10 + 3 = 13
10 + 4 = 14
10 + 5 = 15
10 + 6 = 16
10 + 7 = 17
10 + 8 = 18
10 + 9 = 19
10 + 10 = 20

10 + 2 = 12

10 + 1 = 11

10 + 0 = 10

10 + 5 = 15

10 + 4 = 14

10 + 3 = 13

10 + 8 = 18

10 + 7 = 17

10 + 6 = 16

10 + 10 = 20

10 + 9 = 19

Games and Activities

Games and Activities

I rarely use games or activities other than timed tests to teach addition facts because time in math class is precious. The timed tests suffice to accomplish the objective of all students learning all basic facts. Even the few who don't achieve the mastery standard of 100 facts in three minutes or less by the end of the year know the facts!

Still, games are popular with the students and we do them occasionally "for fun." I try to avoid "Around the World" and similar games because the best students get the most practice, and it is the students having difficulty who *need* the most practice! In this section, I've included three activities which my students enjoy and which allow everyone equal practice.

Addition Grid, p. 85.

Just having students fill in the blanks from memory is a worthwhile activity. As an alternative, the teacher can draw a flashcard and read the *answer* aloud. Students have a short time to find a single "home" for that answer. For example, 15 can be written in the 9 + 6 square, the 8 + 7, etc. A kind of Bingo can be played where students try to place five answers in a row in any direction.

Race to the Finish, p. 86.

This is a favorite activity with students! Directions are on the worksheet on page 86. Allowing students one minute and 15 seconds will approximately equal the speed needed to complete 100 facts in three minutes. Students will fill in 42 answers if they make it completely around the track. Playing "The William Tell Overture" while students are racing makes it even more exciting! This is a good activity for a substitute to do, in lieu of regular timed tests.

Bingo, pp. 87-88.

Twelve different Bingo cards with sums of addition facts are provided. If you'd like to do this activity with fewer facts (0-6, for example), you or the students can make your own cards with graph paper. **As a class activity**, the teacher can draw a flash card and announce the question. Students think of the answer, look for it on their card and cross it out, trying to be first to get a Bingo or blackout. **As a small group activity**, this can be played with a deck of one-sided flash cards (no answers). Four students shuffle the deck, then take turns drawing a card and crossing out the answer on their Bingo cards. If all cards have been used and no one has a Bingo, shuffle and draw again.

Name ____________________________

+	0	1	2	3	4	5	6	7	8	9	10
0											
1											
2											
3											
4											
5											
6											
7											
8											
9											
10											

Race to the Finish!

Directions: Choose any number, 1-10, for your first race. Write that number under "Race 1" at the starting line. Add the number you chose to each of the numbers as you go around the racetrack!

Repeat the race with a different number!

Name: ______________________________

Race 3 | Race 2 | Race 1 | START

FINISH LINE

5 10 0 9 4 3 6 10 4 7 3 8 0 7 4 5 3 10 0 8 2 9 6 1 7 5 2 9 6 1 4 7 2 9 8 5 1 3 6 2 10 1 8

14	13	4	6	11
10	12	18	3	16
20	9	10	15	7
2	5	6	8	19
8	17	11	13	5

5	7	9	13	19
12	6	2	16	8
11	4	10	7	17
20	18	12	14	5
6	15	13	3	9

11	14	19	5	8
9	3	15	12	17
18	7	10	14	6
10	5	9	16	2
15	20	11	4	13

16	9	18	6	13
5	20	3	11	17
7	14	10	8	12
6	2	13	7	15
8	5	19	12	4

16	9	13	3	6
2	15	14	20	12
19	12	10	5	18
14	4	6	11	9
8	10	17	15	7

2	17	6	11	9
5	12	16	4	19
14	7	10	15	5
8	3	18	14	13
15	8	11	7	20

19	2	12	14	3
11	9	15	5	4
5	18	10	8	15
8	11	17	7	12
6	7	20	13	16

15	7	14	3	20
11	2	19	8	13
4	14	10	15	9
12	9	8	17	6
18	5	16	11	8

6	12	7	13	1
20	3	17	5	19
11	16	10	7	12
18	9	2	14	8
8	13	6	15	4

12	7	17	6	11
10	13	18	2	15
3	16	10	9	4
13	9	20	5	14
15	14	8	19	6

7	12	9	14	5
13	9	18	2	7
6	17	10	11	15
20	13	4	6	10
16	8	14	19	3

16	2	6	12	20
11	15	9	17	5
5	6	10	11	13
7	19	14	3	12
14	10	4	18	8

These and other materials are available from Instructional Resources Co.

Materials to Help Students Learn Math

	Quantity	Extended Cost
*Addition Facts in Five Minutes a Day** Item 056 $11.95	____	______
*Subtraction Facts in Five Minutes a Day** Item 064 $11.95	____	______
*Multiplication Facts in Five Minutes a Day** Item 072 $13.95	____	______
*Division Facts in Five Minutes a Day** Item 080 $11.95	____	______
*Casting 9's: A Quick Check for Math Computation** Item 099 $8.00	____	______

Materials to Help Students Learn Reference

	Quantity	Extended Cost
Facts Plus: An Almanac of Essential Information ©1999, 250 pages, ISBN 1-879478-19-6 Item 196 $15.95	____	______
Item 19D, $11.96 each for orders of 25 or more almanacs.	____	______
*Facts Plus Activity Book** ©1995, 182 pages, ISBN 1-879478-11-0 Item 110 $19.95	____	______
Encyclopedia Activity for use with *The World Book Encyclopedia** ©1996, Item 129 $9.50	____	______

Materials to Help Students Learn Spelling

	Quantity	Extended Cost
*Spelling Plus: 1000 Words Toward Spelling Success** Item 048 $19.95	____	______
Dictation Resource Book for use with *Spelling Plus* Item 137 $12.95	____	______
*Homophones Resource Book** Item 145 $15.95	____	______
*Personal Dictionary** Item 153 $3.50	____	______
Spell Well: A One-Year Review for Older Students Item 042 $11.95	____	______

Materials to Help Students Learn Handwriting

	Quantity	Extended Cost
*Manuscript Handwriting Masters** Item 161 $8.50	____	______
*Cursive Handwriting Masters** Item 17X $10.50	____	______

*contains reproducible masters

Total ______________

Shipping ($3.00 for first book, .50 for each additional book) ______________

Total enclosed ______________

Instructional Resources Company
3235 Garland
Wheat Ridge, CO 80033
or
Instructional Resources Company
P.O. Box 111704
Anchorage, AK 99511-1704
907-345-6689 (Phone / FAX)

Name ______________________________

Organization ______________________________

Address ______________________________

____

Phone number ______________________________

❒ Teacher ❒ Home School ❒ Library ❒ Other